MYP Chemist[ry]

A concept-based approach

Years 4&5

Gary Horner

OXFORD
UNIVERSITY PRESS

Acknowledgements

The authors and publisher are grateful to those who have given permission to
reproduce the following extracts and adaptations of copyright material:

American Chemical Society National Historic Chemical Landmarks. 'Flavor
Chemistry Research, USDA ARS Western Regional Research Center,' produced
by the National Historic Chemical Landmarks program of the American
Chemical Society in 2013. Reproduced by permission.

Paul Anastas and John Warner: *Green Chemistry: Theory and Practice* (1998)
Fig.4.1 p.30. www.oup.com. By permission of Oxford University Press.

'Climate Model Suggests Collapse of Atlantic Circulation is Possible' 4 January
2017 from Scripps Institution of Oceanography at UC San Diego. Reproduced
by permission.

Definition of 'Analytical Chemistry' from https://www.nature.com/subjects/
analytical-chemistry. Reproduced by permission.

Definition of 'Analytical Chemistry' from www.acs.org/content/acs/en/careers/
college-to-career, American Chemical Society. Reproduced by permission.

M G Carlin & J R. Dean: Extract from 'Forensic applications of gas
chromatography', 1 January 2013. Reproduced with permission of Taylor and
Francis Group LLC Books via Copyright Clearance Center.

Richard L Clark: Extract from 'Environmentally Friendly Anti-Corrosion
Coatings'. United States Environmental Protection Agency, https://cfpub.epa.
gov. Reproduced by permission.

Extract from 'All About Glaciers,' National Snow and Ice Data Center. Accessed
1 February 2016. https://nsidc.org/cryosphere/glaciers.

Extract from 'California ramps up biofuels infrastructure' from https://www.afdc.
energy.gov 13 October 2016. Reproduced by permission.

Extract from United Nations World Food Programme, www.wfp.org.
Reproduced by permission.

Thomas McCullough, Marissa Curlee: 'Qualitative analysis of cations using
paper chromatography' from *Journal of Chemical Education* Copyright
© 1993, American Chemical Society. Reprinted with permission from the
American Chemical Society.

Matthew Nitch Smith: 'The number of cars worldwide is set to double by
2040', April 4, 2016, *Business Insider Magazine.* , Copyrighted 2016. Business
Insider, 260041:0318PF. Reproduced by permission of Wright's Media.

'Projected Global Transport Growth over the next 25 Years' infographic,
April 4, 2016, *Business Insider Magazine*, © HS, IATA, World Bank, IMF and
Bernstein. Reproduced by permission of AllianceBernstein.

Michael Reilly and Jamie Condliffe: 'A Desert Full of Tomatoes, Thanks to
Solar Power and Seawater', October 6, 2016, *MIT Technology Review Journal*,
Copyrighted 2016. Technology Review. 260318:0418SH. Reproduced by
permission of Wright's Media.

'What is Acid Rain' from United States Environmental Protection Agency,
www.epa.gov/acidrain/what-acid-rain. Reproduced by permission.

XiaoZhi Lim: 'The new breed of cutting-edge catalysts' from 'Nature News',
6 September 2016 published by Nature Publishing Group. Reproduced by
permission of RightsLink.

Contents

How to use this book

To help you get the most of your book, here's an overview of its features.

Concepts, global context and statement of inquiry

The key and related concepts, the global context and the statement of inquiry used in each chapter are clearly listed on the introduction page.

Activities

A range of activities that encourage you to think further about the topics you studied, research these topic and build connections between chemistry and other disciplines.

Worked example

Worked examples take a step-by-step approach to help you translate theory into practice.

Experiments and demonstrations

Practical activities that help you prepare for assessment criteria B & C.

Data-based questions

These questions allow you to test your factual understanding of chemistry, as well as study and analyze data. Data-based questions help you prepare for assessment criteria A, B & C.

ATL Skills

These approaches to learning sections introduce new skills or give you the opportunity to reflect on skills you might already have. They are mapped to the MYP skills clusters and are aimed at supporting you become an independent learner.

 A conceptual question   A debatable question

Summative assessment

There is a summative assessment at the end of each chapter; this is structured in the same way as the eAssessment and covers all four MYP assessment criteria.

Glossary

The glossary contains definitions for all the subject-specific terms emboldened in the index.

Mapping grid

The MYP eAssessment subject list for Chemistry consists of seven broad topics:

Periodic table International Union of Pure and Applied Chemistry The atmosphere

Matter Pure and impure substances Bonding

Types of chemical reaction

These topics are further broken down into sub-topics and the mapping grid below gives you an overview of where these are covered within this book. It also shows you which key concept, global context and statement of inquiry guide the learning in each chapter.

Chapter	Topics covered	Key concept	Global context	Statement of inquiry	ATL skills
1 Balance	Chemical formula Chemical reactions and the conservation of mass Balancing equations Reversible reactions	Relationships	Fairness and development	Imbalanced relationships affect finite resources, both locally and globally.	**Reflective skills:** Develop new skills, techniques and strategies for effective learning **Thinking in context:** Finite resources **Information literacy and communication skills:** Read critically and for comprehension **Thinking in context:** Fairness and development **Thinking in context:** Monocultures and food security **Thinking in context:** Fertilizers vs. pesticides, and CCD
2 Evidence	Metals and non-metals Transition metals Noble gases Fractional distillation of crude oil Alkanes, alkenes, alcohols Atmospheric composition	Relationships	Scientific and technical innovation	Our ability to collect evidence improves with advances in science and technical innovations.	**Critical thinking skills:** Using inductive and deductive reasoning **Critical thinking skills:** Understanding based on new information and evidence **Information and media literacy:** Communicate information and ideas effectively
3 Consequences	Acids and bases Neutral solutions Acid/base reactions, pH and indicators Formation of salts Reactivity series Emissions and environmental implications	Change	Globalization and sustainability	Change as a consequence of human development can be identified within all environments on our planet.	**Critical thinking skills:** Analysing and evaluating issues and ideas **Thinking in context:** Rising sea levels **Thinking in context:** Battery disposal **Thinking in context:** Carbon dioxide emissions

Chapter	Topics covered	Key concept	Global context	Statement of inquiry	ATL skills
4 Energy	Endothermic and exothermic reactions Energy changes in a reaction States and properties of matter Combustion of fuel	Change	Scientific and technical innovation	Scientific and technological advances can enable functional energy transformations within, and between, systems.	Thinking in context: Is our increasing energy usage sustainable?
5 Conditions	Collision theory Chemical reaction kinetics rates Factors affecting rates – temperature, concentration, surface area, catalysts	Systems	Scientific and technical innovation	Scientific innovations advance a scientist's ability to monitor changes in conditions and the effect they have on the rate of a chemical reaction.	Critical thinking skills: Revising your understanding based on new information and evidence Critical thinking skills: Evaluating evidence and arguments Reflective skills: Considering content Research skills: Gathering and organizing relevant information Research skills: Presenting information and data using models and mathematical relationships
6 Form	States and properties of matter Characteristics of gases Solutions, colloids and suspensions Filtration and fractional distillation	Relationships	Identities and relationships	Observing and describing the properties of a substance helps us to understand its identity and how it interacts with the environment.	Communication skills: Organize and depict information logically
7 Function	Formation of salts Filtration, distillation and chromatography The mole concept and chemical calculations	Relationships	Globalization and sustainability	The way in which matter functions is dependent on its properties and the relationship of the different systems within the environment.	Information literacy skills: Finding, interpreting, judging and creating information Information and media literacy skills: Locate, evaluate, synthesize information from a variety of sources Reflection skills: Consider ideas from multiple perspectives
8 Interaction	Redox reactions Corrosion Combustion of fuels Emission and environmental implications Formation of salts Reactivity series	Systems	Globalization and sustainability	The interactions between substances can sometimes be understood and predicted by examining the underlying processes.	Information literacy skills: Process data and report results

Chapter	Topics covered	Key concept	Global context	Statement of inquiry	ATL skills
9 Models	Structure and bonding Electron configuration and valency Properties of elements and compounds Chemical formulae Alloys	Systems	Orientation in space and time	Molecular modelling is used for the visualization of chemical structures, displaying their orientation in space and time.	Creative thinking skills: Apply existing knowledge to generate new ideas, products or processes Thinking in context: Assigning valence electrons of an element using the periodic table Communication skills: Read critically and for comprehension Critical thinking skills: Combine knowledge and understanding to create new perspectives
10 Movement	Redox reactions Reactivity series Electrochemical cell Uses of salts Corrosion Extraction Extraction of metals Diffusion	Change	Scientific and technical innovation	The changes we observe in a chemical system can help us to infer information about the movement of molecules and their properties.	Thinking in context: Utilizing the voltaic cell Information literacy skills: Collect, organize and present information Thinking in context: Electroplating Critical thinking skills: Evaluate evidence
11 Patterns	Periodic trends – groups and periods Atomic structure Electronic configuration and valency Acid and base characteristics	Relationships	Orientation in space and time	Chemists look for patterns in the periodic table in order to discover relationships and trends that help them to predict physical and chemical properties.	Information literacy and communication skills: Communicate information and ideas effectively to multiple audiences
12 Transfer	The mole concept and chemical calculations – concentration Acid–base reactions, pH, titrations and indicators Structural formulae – carboxylic acids and esters	Change	Scientific and technical innovation	Technological advances in analytical devices enhance the ability of scientists to monitor the transfer of matter when changes occur during chemical reactions.	Information literacy skills: Finding, interpreting, judging and creating information

1 Balance

The ozone layer is found in the upper atmosphere, where it acts as a filter preventing over 95% of harmful UV radiation emitted from the Sun from reaching the Earth. Human interactions with the environment may result in the release of substances that can catalyse the breakdown of the ozone molecule. When the balance of the planet's resources and the environment is disrupted, the consequences are experienced globally. What happens when the balance between ozone production and ozone depletion is disturbed?

Fertilizers are easily solubilized in the soil, rapidly breaking down into ammonia. Soluble ammonia captured in rainwater can wash into aquatic environments resulting in an imbalance affecting the natural ecosystems and commercial aquaculture industries. This often results in the rapid growth of algae in ponds and streams. What is the impact of this increased amount of algae?

Camping and making a campfire are enjoyable recreational activities. When wood burns, is there an increase in disorder? Matter undergoes a change of state. Is matter still conserved? Can all matter in this combustion reaction be accounted for?

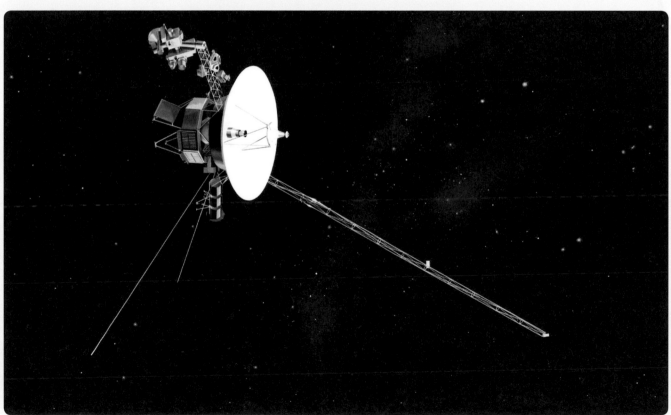

The first law of motion was proposed by Sir Isaac Newton in 1686. Regarded as the definition of inertia, it states that an object will remain at rest or keep moving in a straight line, unless the forces acting on the object become unbalanced due to an external force being applied. Voyager 2 is a deep space probe that was launched by NASA in 1977. It will maintain an approximate speed of 55,000 km h⁻¹ indefinitely unless the forces acting upon it become unbalanced. How has Newton's first law of motion helped us to explore the Universe that we live in?

Key concept: Relationships

Related concept: Balance

Global context: Fairness and development

Introduction

Systems within our universe are dynamic. They constantly undergo both internal and external changes, which require these systems to react and respond. Biological organisms rely on the process of homeostasis to regulate changes that occur within systems and maintain balance. Ecosystems are a dynamic environment of multiple components. Abiotic components include the physical surroundings such as sunlight, water and the soil in which the plants grow. Biotic components include the producers and consumers living within the ecosystem. All of these components contribute to maintaining the balance in conditions and relationships. Uncontrolled development applies external pressures on ecosystems, resulting in an imbalance.

Chemical and physical systems display characteristics that involve a balance in matter and energy, referred to as equilibrium. The control over the balance between reactants and products is essential in many industrial processes and synthetic reactions.

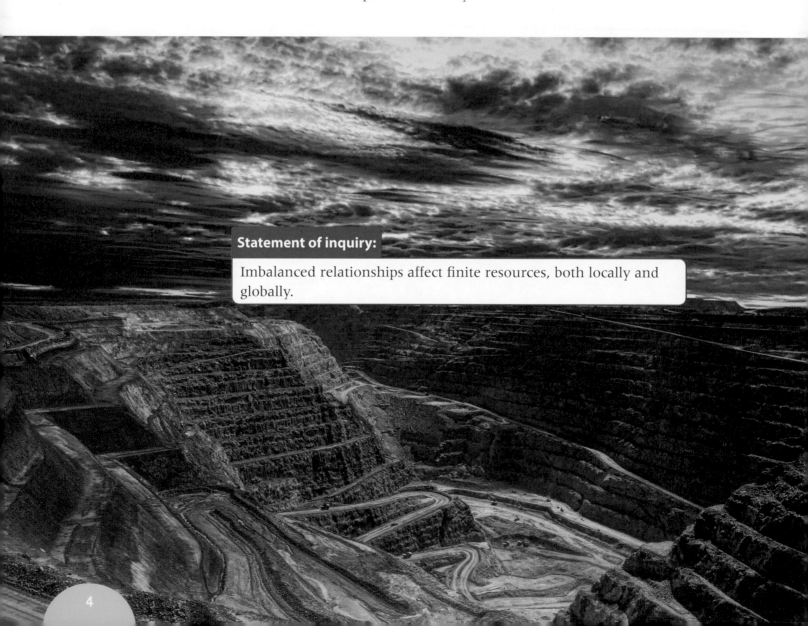

Statement of inquiry:

Imbalanced relationships affect finite resources, both locally and globally.

Is there balance within the universe?

Systems within the universe fall into and out of balance constantly. Imbalanced relationships often have far-reaching effects, so our ability to understand the reasons why a system becomes unbalanced, and in turn rebalanced, is of fundamental importance. In our everyday lives, we too experience changes in balance.

- Biological systems undergo continual change, and to maintain balance is challenging. The diversity of ecosystems, organisms and micro-climates across our planet is immense; and marine environments are some of our planet's largest ecosystems.

- The relationship between organisms and their environments is a delicate balance.

- When excess amounts of nutrients enter a marine ecosystem as the result of increased industrial or agricultural runoff or natural changes in the amounts of available nutrients, this can have a significant and destructive impact.

GENERAL

▲ Harmful algal blooms in marine ecosystems have a major impact within these environments causing an imbalance in the ecosystem. How is aquafarming affected by these occurrences? How might this be a threat to human health?

What are examples of entropy in daily life?

PHYSICAL

Entropy (S) is defined in chemistry as a measure of the distribution of total available energy between particles in a system. The conservation of energy is a fundamental principle of science. When a system has decreasing order and increasing disorder, it is said that its entropy is increasing. However, the total energy within the system remains balanced even when it is distributed in a different way.

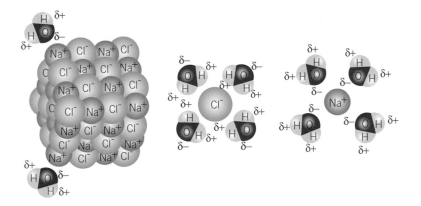

▲ $NaCl(s) + H_2O(l) \rightarrow NaCl(aq)$
When you dissolve sodium chloride in water, the solid ionic compound breaks down into its ions. This change of state from a solid to a liquid is an example of an increase in entropy. The solid lattice structure of sodium chloride is broken down and the ions are free to move in the solution. The amount of disorder increases

ATL Reflective skills

Develop new skills, techniques and strategies for effective learning

Knowledge, skills and understanding are the trilogy of learning. When you graduate from high school, you should aim to take with you knowledge, the skills necessary to acquire knowledge and an understanding of the concepts that you have studied.

Knowledge consists of facts and figures that are ready for you to use. They are often easy to recall and many times are not open to debate. Examples of knowledge in chemistry include:

- The three main states of matter: solids, liquids and gases.

- The formulae of elements and compounds. For example, with oxygen, the element has the symbol O and the formula of the compound oxygen is O_2.

- In a chemical reaction, reactants are on the left-hand side of a chemical equation and products are on the right-hand side.

- Acids are corrosive and have a low pH.

- The main greenhouse gas is carbon dioxide with a chemical formula of CO_2.

Skills are the strategies you develop in order to acquire knowledge and build understanding. The Approaches to Learning skills that we will encounter through this book will help you develop the necessary skills to build your knowledge base and deepen your understanding of scientific concepts.

Understanding is your ability to use your skills and your knowledge, apply them to new contexts and advance your understanding of a concept. It is your understanding of the concepts that enables you to build your knowledge base.

Entropy in your daily life

With the definition of entropy in mind, brainstorm in a small group within your class and identify systems that make up our daily life that have either increasing or decreasing entropy. Justify your choices with supporting arguments. When you have decided on your examples, collectively decide on an effective way to summarize your information so that it can be presented to the other members of the class. Remember that it is important to acknowledge other people's work by creating references and citations in your presentations.

Both languages and symbols are forms of communication that transcend borders, allowing global citizens to communicate with each other. The scientific community uses elemental symbols, formulae and balanced chemical equations to communicate large amounts of information, enabling us to understand what elements and compounds are involved in a chemical reaction.

What can a balanced chemical equation tell us about a reaction?

The study of chemical reactions focuses on substances that are undergoing change. A clear understanding of the energy changes occurring in a chemical reaction is also essential if we are to understand how the reaction occurs under a given set of conditions. The changing balance in energy between reactants and products needs to be understood.

Energy can neither be created nor destroyed, but is converted from one form of energy to another or transferred from one substance to another. The relationship that exists between a system, its surroundings and the universe is well understood (see Chapter 5, Conditions).

The law of conservation of matter states that matter is neither created nor destroyed. Instead, in a chemical reaction, matter is changed from one form to another and can be accounted for at any given time.

All of the resources we use in chemical reactions are in finite supply, but some will be rarer in terms of natural abundance and more expensive. Most reactants in chemical processes are typically less expensive or less finite resources, and are said to be "in excess". It is fundamentally important that we understand the stoichiometric amounts of the reactants required for a given reaction, so that a valuable finite resource is the limiting reactant. A limiting reactant determines the amount of product resulting from a chemical reaction. By designing a method that attempts to completely consume the limiting reagent in the reaction, we minimize wasting valuable resources.

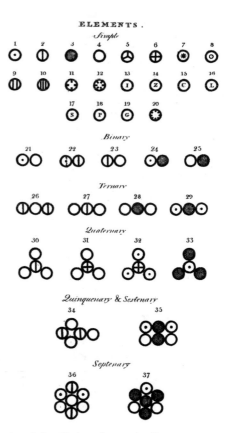

▲ John Dalton's symbolic and visual representations of the atom, published in 1808 in the *New System of Chemical Philosophy*

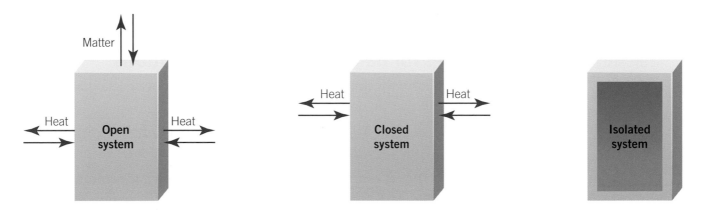

▲ Chemists focus on the system when examining chemical reactions. Matter or energy can flow into and out of an open system altering the balance. This is the more common system being examined. In a closed system, energy can enter and leave the system but the amount of matter remains constant. In an isolated system, both matter and energy cannot enter or leave the system

ATL **Thinking in context**

Finite resources

As economic development throughout the world places demands on the planet's finite resources, we can debate the consequences of connections between fairness and development and finite resources. Your task is to collect and analyze information that enables you to undertake informed discussion with your peers on the following questions:

1. What are some examples of finite resources that are closely managed in an attempt to ensure fair and equitable distribution?

2. Who decides on the distribution of resources within communities and between all living things?

3. What structures are in place within societies to monitor and protect the reserves of finite resources?

4 Should some societies be able to consume more finite resources than others?

QUANTITATIVE

How do chemists manage finite resources in chemical reactions?

Mathematics, the language of science, communicates that a balanced chemical equation is a proportional relationship between reactants and products. At the atomic level, individual atoms of elements are reacting with each other to form new compounds. How do we make sense of this at the macroscopic level?

The reactions between elements and compounds, and the chemical and physical properties of reactants and products are the focus of chemical research. Many different types of chemical reactions exist, such as combustion, single replacement, synthesis, decomposition and reduction-oxidation. In simple terms, a chemical reaction is a rearrangement of atoms, involving the breaking and making of bonds in the reactants and products. How then do we balance a chemical equation?

Rules for balancing chemical equations

1. Balance the atoms of the metal element on the reactant and product side of the reaction first.

2. Balance the atoms of any element found in only one chemical formula on either side of the reaction.

3. Balance the atoms of any remaining elements, if required.

4. Check to see that there are the same number of atoms of each element on both sides of the chemical equation.

▲ With a single proton, hydrogen is the simplest element that exists. This is the first image ever recorded of the orbital structure of a hydrogen atom. The instrument used to record the image is called a quantum microscope

Transition metal nomenclature

When metallic elements have more than one oxidation state, such as transition elements, Roman numerals are used to represent the oxidation number. For example, iron(III), copper(II), manganese(VII).

Worked example: Balancing equations

Question

Metallic iron rusts in the presence of oxygen and water, losing three electrons to form the iron(III) cation. This is an oxidation reaction impacting steel and iron structures such as automobiles and bridges, causing hundreds of millions of dollars in damages each year.

$$\text{iron} + \text{oxygen} \rightarrow \text{iron(III) oxide}$$
$$Fe(s) + O_2(g) \rightarrow Fe_2O_3(s)$$

Balance the equation shown.

Answer

First, we need to balance the metal element, in this case iron. There is one iron atom on the reactant side and two iron atoms on the product side of the equation. Add a coefficient of 2 on the reactant side to the element iron: the metal element is balanced.

$$2Fe(s) + O_2(g) \rightarrow Fe_2O_3(s)$$

Then we need to balance elements found in only one chemical formula. Oxygen is only found in one chemical formula, molecular oxygen, on the left side of the equation. There are an even number of oxygen atoms on the reactant side and an odd number of oxygen atoms on the product side.

In mathematics, when adding fractions that have a different denominator, you calculate the lowest common denominator (LCD).

The LCD of 2 and 3 is 6. The number of atoms of oxygen on the reactant and product side must be 6.

$$2Fe(s) + 3O_2(g) \rightarrow 2Fe_2O_3(s)$$

Then, we need to balance any remaining elements and perform a final check to see whether there are the same number of atoms of each element on both sides of the equation. There are not any remaining elements we haven't looked at, but the previous step has imbalanced the number of iron atoms. We therefore need to double the coefficient for the iron atoms on the reactant side.

$$4Fe(s) + 3O_2(g) \rightarrow 2Fe_2O_3(s)$$

1. Fertilizers are produced by a reaction between an acid and ammonia. The following word equations summarize the formation of a variety of fertilizers. Write balanced chemical equations for these, including the states of matter for all reactants and products.

   a) Ammonia + nitric acid → ammonium nitrate

   b) Ammonia + sulfuric acid → ammonium sulfate

   c) Ammonia + phosphoric acid → ammonium phosphate

   d) Potassium hydroxide + nitric acid → potassium nitrate + water

2. The combustion of methane gas is an important source of energy. Methane is the most abundant compound in natural gas. Balance the chemical equation for the combustion of methane.

$$CH_4(g) + O_2(g) \rightarrow CO_2(g) + H_2O(g)$$

3. Aluminium and iron are both common metals used in the manufacturing and construction industries. While iron is easily oxidized in the presence of water and oxygen, known as rusting, aluminium forms an oxide layer that prevents further oxidation. Balance the following chemical equations:

   a) $Al(s) + O_2(g) \rightarrow Al_2O_3(s)$

   b) $Fe(s) + O_2(g) \rightarrow Fe_2O_3(s)$

(A)(B)(C)(D) Experiment

Testing for the copper(II), iron(II) and iron(III) ions

There are two standard analytical tests for the presence of copper(II), iron(II) and iron(III) ions in a solution, one involving the addition of sodium hydroxide and the other the addition of ammonia solution. The formation of insoluble metal hydroxides leads to the identification of these ions, as characteristic color changes can be observed.

Materials

- 0.5 mol dm^{-3} copper(II) nitrate
- 0.5 mol dm^{-3} iron(II) nitrate
- 0.5 mol dm^{-3} iron(III) nitrate
- 1.0 mol dm^{-3} ammonium nitrate
- 1.0 mol dm^{-3} sodium hydroxide
- Test tubes and rack
- Dropping pipettes

Method and questions

1. Using a dropping pipette, transfer 3–5 cm^3 of 0.5 mol dm^{-3} copper(II) nitrate solution into a test tube and observe the color.

2. Then, using a different pipette, add 1.0 mol dm^{-3} sodium hydroxide solution in a dropwise manner. Record your observations in the table below.

3. Continue adding the sodium hydroxide solution and observe any changes that are taking place. Record your observations in the table below.

Metal cation	Initial observations	Final observations (sodium hydroxide reaction)
Cu^{2+}		
Fe^{2+}		
Fe^{3+}		

4. Repeat steps 1–3 with fresh solutions of each of the metal ion solutions.

5. Repeat steps 1–3 with fresh solutions of each of the metal ion solutions and 1.0 mol dm^{-3} ammonium nitrate solution. Create a similar table to record your observations when excess ammonium nitrate is used.

Read critically and for comprehension

In 1998, recognizing that our planet has a finite amount of resources and that the pace of industrialization and consumer-based economies was having a negative impact on the environment, Paul Anastas and John Warner developed a series of twelve guiding principles for chemical industries and manufacturing to follow when performing chemical synthesis and other chemical reactions. These principles are:

1. Prevent waste
2. Use of renewable feedstock
3. Atom economy
4. Reduce derivatives
5. Less hazardous waste
6. Catalysts
7. Design benign chemicals
8. Design for degradation
9. Benign solvents and auxiliaries
10. Real-time analysis for pollution prevention
11. Design for energy efficiency
12. Inherently benign chemistry for accident prevention

Anastas and Warner held the belief that there was a need for industry to reduce the amount of waste, both hazardous and non-hazardous, by devising methods that would create fewer unwanted products. Some of their proposed methods include: using less toxic solvents which could harm the environment, using renewable materials where possible (rather than diminishing the amount of finite resources)*, and using methodologics and building industrial infrastructure to prevent the loss of energy and recycle excess energy. Their initiatives have been embraced by many sectors of industry and governments alike, and in some countries the principles of green chemistry have been transformed into laws that govern the use of chemical technology.

Discuss the merits, and the large scale applicability, of green chemistry principles.

Research some examples of how green chemistry has been incorporated into existing industrial manufacturing systems to increase efficiency and decrease environmental effects. Some questions to consider include:

1. What are some characteristics of balanced industrial manufacturing systems?
2. How can imbalanced relationships produce both local and global effects?

*The law of conservation of matter states that matter can be neither created nor destroyed; however, non-renewable resources are considered to be finite as a consequence of the time required for them to form. As non-renewable resources are transformed into different compounds, the atoms are rearranged rather than lost. Discuss why some people within the international community are concerned about the decrease in availability of non-renewable resources.

What is a chemical equilibrium?

GENERAL

Many systems, including chemical reactions, exist in a state of equilibrium. An example of this is the hemoglobin-oxygen interaction: hemoglobin is the iron-containing protein in red blood cells, which is responsible for transporting oxygen to cells. Each hemoglobin molecule can carry up to four oxygen atoms, and the equilibrium reaction can be expressed as follows (where "Hb" represents hemoglobin):

$$Hb(aq) + 4O_2(g) \rightleftharpoons Hb(O_2)_4(aq)$$

This equilibrium is maintained as long as sufficient oxygen is supplied.

▶ Many fish have an organ called a swim bladder. This increases their ability to control buoyancy and therefore more easily maintain a preferred water depth without constantly swimming. These fish can be considered to be in balance with their environment

ATL Thinking in context

Fairness and development

In some cases, social and economic development throughout the world has led to greater control of the planet's finite resources. We therefore can consider some possible connections to fairness and development with respect to these finite resources. With this in mind, discuss the following questions in small groups:

1 To what extent should scientists consider moral and ethical obligations when conducting their research?

2 When scientific research and discoveries have the potential to disrupt balance, to what extent are governments responsible for safeguarding society from this research and these discoveries?

GENERAL

What impact did the Haber process have on our planet?

> Gaseous nitrogen combines with gaseous hydrogen in simple quantitative proportions to produce gaseous ammonia.
>
> **Fritz Haber**

German chemist Fritz Haber was awarded the Nobel Prize in Chemistry in 1918 for developing a method to synthetically fix nitrogen from the air. Before Haber's successful breakthrough, science had been unable to synthetically replicate this biological process.

Haber's discovery allowed for the large scale production of fertilizers that began during the Green Revolution and continues today. His process also provided Germany with a source of ammonia that was used for the production of explosives during World War I. The Green Revolution began in the mid-20th century, when advances

The large scale, industrial production of ammonia is an example of a once finite resource becoming much more easily controlled

in chemical fertilizers, synthetic pesticides and herbicides changed the face of agriculture forever. Advances brought a new method of growing crops called high-crop yield, which improved the amount of food crops being produced and available to feed a growing global population. The Green Revolution, with the help of science and technology, continues today.

What do you envision when you hear the term "chemical equilibrium"? Perhaps you consider that a better understanding of a chemical reaction and the optimal conditions for it to occur will enable you to better control the reaction. Industrial chemists understand that in order to maximize their product yield, they have to analyze how reactions work and how efficiency within these reactions can be increased. This can often involve the manipulation of finite resources.

In economic terms, equilibrium occurs when supply and demand are equal. In a chemical system at equilibrium, the forward and reverse reaction rates are equal. No changes in the macroscopic properties of the reaction are observed when a system has reached equilibrium. The position of the equilibrium is dependent on the chemical reaction and the conditions under which it is taking place. Some chemical reactions favour the product side, while other chemical reactions favour the reactant side.

Equilibrium is linked to a stable balance where objects and systems are static. In chemical systems, reactions at equilibrium are dynamic but the concentrations of reactants and products remain constant

How has industrial-scaled ammonia production impacted the global population?

At the beginning of the 20th century when the population of the world was 1.6 billion people, a shortage of synthetically produced fertilizers and a growing demand for increased food production encouraged Fritz Haber to develop a method that would enable the large scale production of ammonia. As ammonia is an essential component used to provide nitrogen for crop fertilizers, today the annual global production of ammonia stands at over 150 million tonnes and rising. Ammonia is mainly converted into other important compounds such as nitric acid, HNO_3, ammonium nitrate, NH_4NO_3, ammonium sulfate, $(NH_4)_2SO_4$, and urea, NH_2CONH_2.

The Haber process involves the reaction of 1 mol of nitrogen gas and 3 moles of hydrogen gas in the presence of a catalyst:

$$N_2(g) + 3H_2(g) \rightleftharpoons 2NH_3(g) \qquad \Delta H = -92 \text{ kJ mol}^{-1}$$

Haber's genius was to investigate and propose the ideal conditions under which the highest yield possible would be achieved in the shortest amount of time. He did this by determining the ideal temperature and pressure for the reaction.

How do changes in reaction conditions affect an equilibrium system?

A number of very significant industrial processes, including the production of ethanol and ammonia, are reversible reactions and can be explained by the application of **Le Chatelier's principle.** This principle states that if a system in equilibrium is disturbed by changes in temperature, pressure or concentration of the components, the system acts to oppose the change and rebalance the system.

This principle is named after Henry Louis Le Chatelier, a French chemist whose pioneering work opened the door to a better understanding of how equilibrium reactions behave. As we will see, this principle has critical implications for the production of ammonia (once considered a finite resource) and its use in large-scale agricultural applications.

In 1901, Le Chatelier unsuccessfully attempted nitrogen fixation in his laboratory, resulting in a massive explosion. The process was later perfected by Fritz Haber

According to Le Chatelier's principle, when a change is made to the conditions of a chemical equilibrium, the position of the equilibrium will readjust to minimise the change made. The balance between the forward and reverse reactions will shift to offset the change and return the system to equilibrium.

At equilibrium:

- The forward and reverse reactions are occurring at equal rates and no overall changes are produced.

- The concentrations of reactants and products remain constant.

- Within the reactant/product mixture, there is no change in the observable (macroscopic) properties such as color and density.

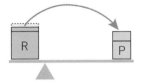

This reaction is in equilibrium. The concentration of reactants is twice that of the products.

More reactants have been added. The equilibrium has been disturbed.

The equilibrium shifts to the right. Reactants are changed into products until the equilibrium is restored. The relative concentrations of reactants and products are the same as before.

ATL Thinking in context

Monocultures and food security

The use of synthetic fertilizers is a predominant global trend in agriculture. The Green Revolution allows, for the first time in mankind's history, for food production to be in excess of the increasing demand. However, the monocultures that dominate the agricultural landscape today are often viewed as unsustainable. While the productivity of annual food crops is of vital importance for feeding an ever growing world population, it can be argued that the prevalence of monocultures that dominate the global agricultural landscape is actually a threat to food security, the continuation of species and the conservation of natural resources throughout our planet.

According to the United Nations World Food Programme: "People are considered food secure when they have available access at all times to sufficient, safe, nutritious food to maintain a healthy and active life."

Research some of the threats posed by the monoculture model, considering both threats to food security and to the conservation of natural resources. Some questions you can consider include:

1. What are some characteristics of balanced agricultural systems with regard to crop selection and crop rotation?

2. What are some of the possible alternatives to monocultures?

3. How can imbalanced relationships produce both local and global effects?

4. How are imbalances in traditional weather patterns and climate trends affecting agricultural communities around the world?

What changes will affect the equilibrium position in the Haber process?

Concentration of reactants or products

According to Le Chatelier's principle, at a given temperature, the position of the equilibrium will change in response to a change in concentration of reactants or products. When industry collects ammonia by changing the pressure, thereby decreasing the concentration of ammonia present in the reaction vessel, the equilibrium will change to counteract this change.

When the concentration of one or more reactants is increased, the forward reaction will be favoured to minimize this change. Effectively, the reaction will try to use up the additional reactants. By decreasing the concentration of a product, the forward reaction will also be favoured to replenish the missing products. Both of these changes in concentration help to force the reaction in the forward direction and manufacture more ammonia.

Chemists and chemical engineers who work at ammonia production facilities understand the principles of the Haber process and adjust reaction conditions to maximize yield and in turn profits.

Pressure

The Haber process is an equilibrium system that involves only substances in the gaseous state. This means that the system will be affected by changes in applied pressure. In this reaction, the number of moles of gas on the reactant and product side is unequal. How many moles of gas are on the reactant side and how many are on the product side?

A change in pressure applied to the system will result in a shift in the equilibrium position, either favouring the forward or reverse reactions.

Consider the fact that 1 mol of a gas under standard conditions (temperature of 273 K and pressure of 1 atm) occupies 22.7 dm³ of volume.

According to Le Chatelier's principle, an increase in pressure applied to the system will favour the side of the reaction with the least number of moles of a gas. For example, in the case of nitrogen gas reacting with hydrogen gas, the forward reaction is favoured and the yield increases:

$$N_2 + H_2 + H_2 + H_2 \rightleftharpoons NH_3 + NH_3$$

Temperature

The production of ammonia is an **exothermic** process, where energy is released as heat into the surroundings during the reaction. The reverse reaction, which produces hydrogen and nitrogen gas, is therefore **endothermic**, and takes in energy from the surroundings. When the system is at equilibrium, the forward and reverse reactions

> Gas molecules are normally far apart and moving at high velocities. By lowering the temperature and increasing the pressure of a closed system, gas molecules move closer to each other and eventually undergo a change of state from gas to liquid. This process is known as liquefaction.

▲ An industrial ammonia production plant

occur at equal rates and there is no change in energy. Increasing the temperature of a reaction effectively adds additional reactant for an endothermic reaction, or product for an exothermic reaction. Therefore in the Haber process, increasing the temperature is adding to the ammonia side, so the reverse reaction is favoured to balance the equilibrium.

- What has happened to the yield when the temperature is increased?

- What is the advantage of increasing the temperature of the system?

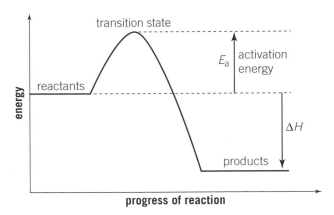

progress of reaction

▲ How does the total amount of energy contained in the reactants compare to the products? Which substances will be more energetically stable, reactants or products?

 Experiment

Making fertilizer

 Safety

- Wear eye protection

- Sulfuric acid is corrosive—avoid contact with the skin

- Ammonia is a corrosive gas to the respiratory system—perform the experiment in a well ventilated room or fume hood.

Materials

- 1.0 mol dm⁻³ sulfuric acid
- 1.0 mol dm⁻³ ammonia solution
- 25 cm² measuring cylinder
- Porcelain evaporating basin
- Universal indicator paper
- Bunsen burner

Method

1. Measure 20 cm² of 1.0 mol dm⁻³ sulfuric acid with a measuring cylinder and transfer it into a porcelain evaporating basin. Place the basin on top of a wire gauze and tripod.

2. With a dropping pipette, slowly add small amounts of 1.0 mol dm⁻³ ammonia solution to the basin until there is a constant smell of ammonia coming from the mixture.

3. After each small addition, check the pH of the solution by stirring the mixture with a glass stirring rod and placing a drop of the liquid onto a piece of universal indicator paper.

4. When the pH has moved above 7.0, sufficient ammonia has been added.

5. Over a moderate Bunsen burner flame, warm the solution until it has reduced in volume to about 20% of the original. Care should be taken not to allow the solution to be heated too strongly as the solution may spit.

6. When crystals begin to form in the basin, cool, filter and allow the crystals to dry.

Questions

1. Write the word and balanced chemical equation for the reaction.

2. Which element found in the product is responsible for plant growth? Explain your answer.

Research

1. What is the source of the reactants, nitrogen and hydrogen, used in the Haber process?

2. Write a balanced chemical equation for the production of hydrogen gas.

Uses of nitrogen-containing chemicals

Gather information from a variety of sources about the main uses of the chemicals nitric acid, HNO_3, ammonium nitrate, NH_4NO_3, ammonium sulfate, $(NH_4)_2SO_4$ and urea, NH_2CONH_2. Structure the information as a summary and present it. Based on your research, share ideas with your peers, while answering the following question: Why must the total amount of ammonia production keep increasing each year?

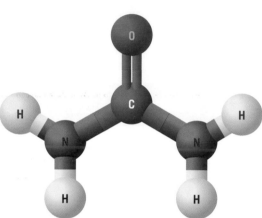

▲ Urea is a commonly used component of fertilizers and an integral part of the agricultural industry

Data-based question: Industrial reaction conditions for the production of ammonia

1. What temperature achieves the highest yield of ammonia? [1]

2. Describe how the yield changes as the pressure is increased. [1]

3. The ideal temperature used by Haber is approximately 450°C. Outline the reasons why a higher temperature would be used to increase production of ammonia instead of a lower temperature. [2]

4. A pressure of 200 atm is used during the process. Why does industry not use a much higher pressure to maximise yield? [2]

5. Explain with reference to the position of the equilibrium why increasing the pressure of this closed system favours the forward reaction. [2]

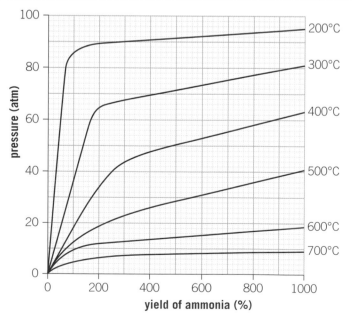

Fertilizers vs. pesticides, and CCD

Fertilizers and pesticides are both classes of synthetic compounds used in the agricultural industry; while fertilizers are typically nitrogen-based and applied to increase plant growth/yield, pesticides are applied to control insects considered to be crop pests. Imidacloprid is a pesticide belonging to the class of chemicals called neonicotinoids, which destroys insects by producing fatal effects within their central nervous system.

▲ Chemical structure of imidacloprid

Both fertilizers and pesticides have been documented to disrupt balance within environments. Of particular concern are recently documented links between honey bee exposure to imidacloprid and an increase in the rate of colony collapse disorder (CCD). In January 2013 the European Food Safety Authority stated that the use of imidacloprid presents an unacceptably high risk to honey bees and that the industry-sponsored scientific research and findings, upon which claims of safety have relied upon, may be flawed.

Research some of the current data related to colony collapse disorder (causes, effects, where it is occurring, solutions), and what the implications to our global food supply will be if pollinators continue to decrease in numbers on a large scale. Some questions to consider include:

1. Might pollinators be considered a finite resource? Why or why not?

2. What are some characteristics of balanced agricultural systems with regard to pollinators?

3. When scientific discoveries disrupt balance (for example, the design and application of imidacloprid), to what extent are governments responsible for safeguarding society from the harmful effects?

4. How are imbalances in the number of pollinators and beekeepers affecting agricultural communities around the world?

5. How can imbalanced relationships produce both local and global effects?

Experiment

Decomposition of ammonium chloride

⚠ **Safety**

- Wear eye protection
- Perform this reaction in a well ventilated area

Materials
- Solid ammonium chloride
- Bunsen burner and tongs
- Test tube

Pre-investigation question

How do you test for the presence of ammonia gas?

$NH_4Cl\ (s) \rightleftharpoons NH_3\ (g) + HCl\ (g)$

 The decomposition of ammonium chloride is an endothermic reaction which demonstrates the reversible nature of some chemical reactions

Method

1. Place 1–2 cm of ammonium chloride powder into the bottom of a dry test tube.

2. Light the Bunsen burner and slowly pass the test tube through a strong flame until you observe evidence that the reaction has started. At this point, stop heating the test tube.

Questions

1. What do you observe when the test tube of ammonium chloride is heated?

2. A white solid is observed at the top of the test tube? Can you identify the substance?

3. Explain why this solid has re-formed away from the source of heat.

Demonstration

Testing for the ammonium ion

⚠ Safety

- Wear eye protection

- Reaction must be performed in the fume hood

- Concentrated hydrochloric acid is a highly corrosive substance and should not be handled by students

- Ammonia gas is harmful to the respiratory system when inhaled

Materials

- 1 mol dm⁻³ sodium hydroxide

- Solid ammonium chloride salt

- 6 mol dm⁻³ hydrochloric acid

- Red litmus paper

- Test tube and tube rack

- Dropping pipette

- Spatula

- Bunsen burner

Method

1. Using a dropping pipette, transfer 2 cm³ of 1 mol dm⁻³ sodium hydroxide into a test tube.

2. Add 1 flat spatula of solid ammonium chloride salt.

3. Gently warm the contents of the test tube by passing it back and forth through a Bunsen burner flame.

4. Replace the test tube in the test tube rack and hold a piece of damp red litmus paper at the rim of the test tube. Record any observations.

5. Dip a stirring rod into concentrated hydrochloric acid (corrosive) and place the stirring rod at the tip of the test tube. What do you observe?

Questions

1. Name some other compounds that react with damp red litmus paper. What do they all have in common?

2. Construct a balanced chemical equation to describe the reaction that you observed when you heated the mixture.

3. Give the name and formula of the product of the reaction between ammonia and concentrated hydrochloric acid.

un-ionized   ionized

◀ Ammonia gas, NH$_3$, is a noxious gas with a strong pungent smell. Ammonia is often confused with the ammonium ion, NH$_4^+$

How are equilibrium systems controlled in industrial processes?

Each year, the Contact process is used to produce over 200 million tonnes of sulfuric acid. Over 50% of total global production is used to manufacture phosphoric acid, H$_3$PO$_4$. Important products from the reactions of sulfuric acid and phosphoric acid include the fertilizers ammonium sulfate and ammonium phosphate. Sulfuric acid is used in many other processes including the manufacture of plastics, batteries and paints.

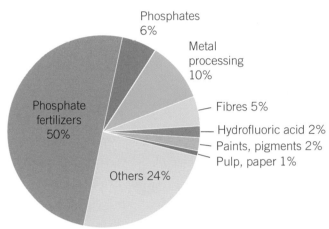

▲ The uses of sulfuric acid

The Contact process is the conversion of sulfur dioxide into sulfur trioxide:

$$2SO_2(g) + O_2(g) \rightleftharpoons 2SO_3(g) \qquad \Delta H° = -196\,kJ\,mol^{-1}$$

This process again utilises Le Chatelier's principle. The reactant side of this chemical reaction contains 3 moles of gases while the product side contains 2 moles of a gas. By raising the pressure of this system, the equilibrium will move to reduce the pressure and the forward reaction will be favoured. This will increase the yield of sulfur trioxide. Catalysts are also added to the reaction mixture to maximise the rate of production.

The Ostwald process: nitric acid production

The Ostwald process is another industrial application of Le Chatelier's principle. This process is used for making nitric acid and involves the catalytic oxidation of ammonia (Haber process), as the first step.

Research the oxidation of ammonia through the Ostwald process, and respond to the following questions:

1. What is oxidation?
2. Explain the equilibrium process involved and the conditions used in this reaction.
3. What is nitric acid used for?
4. State the role of nitric acid in this industrial process. Give some other examples of catalysts being used in industrial processes.

Summative assessment

Statement of inquiry:

Imbalanced relationships affect finite resources, both locally and globally.

Introduction

In this summative assessment, you will first look at how Le Chatelier's principle can be used to predict the effect of changes on the equilibrium position of chemical reactions. You will then design a simple experiment to demonstrate how changing reaction conditions can alter the chemical equilibrium. Next, you will analyse experimental data and suggest what changes in reaction conditions have taken place. Finally, you will look the impact large-scale fertilizer production and usage is having on the global community.

Le Chatelier's principle

Le Chatelier's principle states that if a system in equilibrium is disturbed by changes in temperature, pressure, and/or concentration of the components, the system will shift the position of the equilibrium to counteract the change and return the system to balance.

For each of the following chemical reactions, examine the reactants and products and their states of matter, and decide how the change in reaction conditions will affect the position of the equilibrium.

1. The Haber process describes the industrial production of ammonia gas on a large scale:

$$N_2(g) + 3H_2(g) \rightleftharpoons 2NH_3(g) \qquad \Delta H° = -92 \text{ kJ mol}^{-1}$$

Predict the effect of the following changes on the position of the equilibrium in the Haber process.

   a) Nitrogen gas is added to the system at equilibrium. [1]

   b) Hydrogen gas is removed from the system at equilibrium. [1]

   c) The pressure of the system is decreased. [2]

   d) The temperature of the system is increased. [2]

2. The decomposition of sulfur trioxide is an endothermic process:

$$2SO_3(g) \rightleftharpoons 2SO_2(g) + O_2(g) \qquad \Delta H° = +196 \text{ kJ mol}^{-1}$$

Predict the effect of the following changes on the position of the equilibrium:

a) Sulfur trioxide gas is removed from the system at equilibrium. [1]

b) Oxygen gas is removed from the system at equilibrium. [1]

c) The pressure of the system is increased. [2]

d) The temperature of the system is decreased. [2]

Exploring changes in equilibrium reactions

When the cobalt(II) chloride, a blue transition metal complex, is dissolved in water, it establishes an equilibrium with the pink colored cobalt(II) hexahydrate ion. The forward reaction in this equilibrium is exothermic.

$$[CoCl_4]^{2-}(s) + 6H_2O(l) \rightleftharpoons [Co(H_2O)_6]^{2+}(aq) + 4Cl^-(aq)$$
$$\Delta H^0 = -X \text{ kJ mol}^{-1}$$

3. Design an experiment that enables you to observe how the equilibrium of this system can be changed using the following materials: cobalt(II) chloride hexahydrate powder, 2 mol dm^{-3} hydrochloric acid and distilled water.

Your design should include the following features:

- a method (including all apparatus) used to establish the initial equilibrium

- two different methods used to demonstrate how the equilibrium position can be altered

- an appropriate way of recording both quantitative and qualitative observations of the reactions you are performing

- details of the variables being controlled and evidence of an awareness of safety issues. [10]

How does chemical bonding relate to electrical conductivity?

The equilibrium that exists between nitrogen(IV) oxide and dinitrogen tetroxide can be monitored by observing the color change with changing temperature:

$$2NO_2(g) \rightleftharpoons N_2O_4(g)$$

NO_2 is a brown colored gas and N_2O_4 is colorless.

Some students performed a series of simple experiments to investigate how they could change the position of the equilibrium. They recorded their qualitative observations in a data table.

Reaction	Initial color at equilibrium	Final color
A	Light brown	Color lightens
B	Light brown	Color darkens
C	Light brown	Color darkens
D	Light brown	Color lightens

Interpret the data and by applying your understanding of Le Chatelier's Principle, answer the following questions:

4. **a)** Suggest one or more changes in reaction conditions that may have taken place at room temperature resulting in the observations for:

   i) Reaction A [4]

   ii) Reaction B [4]

   b) Reaction C underwent changes in the temperature of the reaction mixture. The reaction mixture was placed in a water bath at a temperature of 45°C. Predict if the reaction is endothermic or exothermic. Give reasons to support your decision. [4]

   c) The reaction mixture for D was placed in a bath of ice-water. Analyze the results of reaction D and comment whether these results support or contradict the decisions you made about reaction C. [3]

5. Choose any of the reactions and describe the changes you would observe, if any, on the addition of a catalyst to the reaction mixture. [2]

6. Suggest a possible extension to this investigation. [3]

 Impacts of the availability of fertilizers

The Haber-Bosch process is considered by many to be one of the most important transformational technological developments of the modern age, with some of the most substantial unintended consequences.

Choose one of the research tasks below and write a report as instructed.

7. A significant environmental effect of the widespread use of agricultural fertilizers on our planet is eutrophication: the accelerated growth of aquatic plants and algae in bodies of water that contain excess nutrients (namely, nitrogen and phosphorous) as a result of agricultural runoff. This increased presence of plants and algae depletes the levels of dissolved oxygen in the water, referred to as hypoxia, and often leads to areas of open water that are unable to sustain life (also known as "dead zones").

As a chemical hydrologist specializing in nutrient loss strategies, you are producing a community report on the health of local watersheds. Research the specific environmental impacts and scientifically viable prevention solutions related to agricultural runoff in a river of your choice. Your report can be written, visual, or a combination of the two, and must include:

- how the watershed has changed over the last 15-20 years

- current measures that are being implemented to restore health to the watershed

- a justified proposal for future measures that can be taken to reduce or eliminate nonpoint source pollutants.

Moreover, your report should also answer the following questions:

a) How can imbalanced relationships produce both local and global effects?

b) What are some of the consequences (moral, ethical, and/or environmental) of infinite resources becoming finite? [20]

or

8. Another significant effect of the widespread use of agricultural fertilizers on our planet has been a skyrocketing 20th century global population, which increased from 1.6 billion to 6 billion. At the time of print nearly two decades into the 21st century, the global population is over 7.5 billion and continues to rise.

You are an ambassador with the United Nations Environment Programme working on the impact of applied nitrogen in agriculture, and advocating for the UNEP proposed "20:20 for 2020" target. The target aims to improve the efficiency use by 20 percent and to reduce its overall use by 20 million tons each year up to 2020.

Write a report addressing the following points:

- how and why population models have changed over the past century

- current population forecasts for the next century

- implications of an ever-increasing global population

- a summary of proposed measures to reduce the application of nitrogen based fertilizers in agriculture.

Your report should also respond to the questions:

a) How can imbalanced relationships produce both local and global effects?

b) What are some of the consequences (moral, ethical, and/or environmental) of finite resources becoming infinite? [20]

2 Evidence

> **"**
>
> **Science is a way of thinking much more than it is a body of knowledge.**
>
> Carl Sagan
>
> **"**

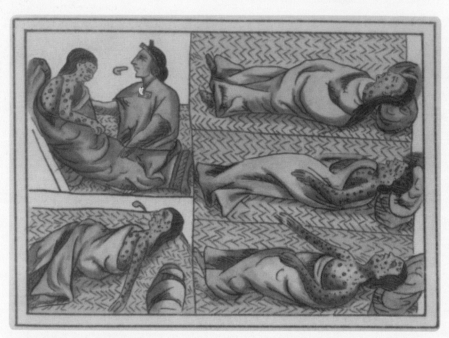

▲ In 1520, Spanish forces landed in what is now Veracruz, Mexico, and introduced smallpox. This resulted in the death of almost half the population of the capital of the Aztec Empire. Smallpox is a contagious infectious disease. In the late 18th century it was observed that milkmaids seemed to be immune to smallpox and this was attributed to the fact that many had previously had cowpox (a disease similar to smallpox, but much less dangerous). This observation led to the development of effective vaccines. In what other ways is gathering evidence important in medicine?

▲ Galileo discovered Jupiter with this telescope. He saw four very bright bodies close to Jupiter that he initially thought were stars. However, with continued observations he realized that their positions relative to Jupiter changed and sometimes they disappeared. The only explanation was that they were moons orbiting Jupiter, disproving the belief that all celestial bodies rotated about the Earth. How have scientists gathered evidence about our universe?

▲ Drones can be used to gather evidence about the growth of crops. The information gathered can also be used to deliver fertilizer to the crops as required. What advantages does this technology offer compared to traditional methods?

Kenai Fjords National Park
Exit Glacier Was Here

If you had been standing here in 1998 the ice would have been right at your feet. The trail in front of you traverses a landscape recently exposed as this side of Exit Glacier melted back.

As you walk the trail, watch for clues of life rebounding in the wake of the glacier. Crusty lichens and mosses crumble rocks into new soil, and tiny seedlings struggle to gain a toehold. Can you image how this place will look in 10 or 20 years?

◀ This photo of Alaska's Exit Glacier was taken in 2016. The sign shows the position of the glacier in 1998. Scientists gather evidence about changes in the size and position of glaciers. This evidence contributes to our understanding of the impacts of global warming. What are the consequences of glaciers melting?

Key concept: Relationships

Related concept: Evidence

Global context: Scientific and technical innovation

Introduction

The scientific method is a collection of approaches to experimentation, designed to increase our knowledge and understanding of scientific concepts, and enable us to test the validity of our conclusions. Observations, both qualitative and quantitative, can prompt a question or identify a problem. From this a hypothesis can be formed; this is a statement that predicts the outcome of future observations. The next stage is to design an experiment to test the hypothesis. An important part of this is to identify the variables that you want to control and the variables you want to measure. You also need to think about ways to make your evidence as reliable as possible, for example, by repeating your measurements.

The next stage is to gather evidence—in other words perform the experiment. It is important to record and retain all your data. It may be tempting to discard data that doesn't fit your preconceptions, but many discoveries have arisen from results that initially "looked wrong". The evidence is then analyzed to determine how it supports or disproves the hypothesis.

▼ Advances in technology and the refinement of instrumentation have allowed scientists to collect data that was previously impossible to access. The accuracy of DNA profiling enables us to trace ancient human migration patterns and the origin of human populations. How else is DNA profiling used?

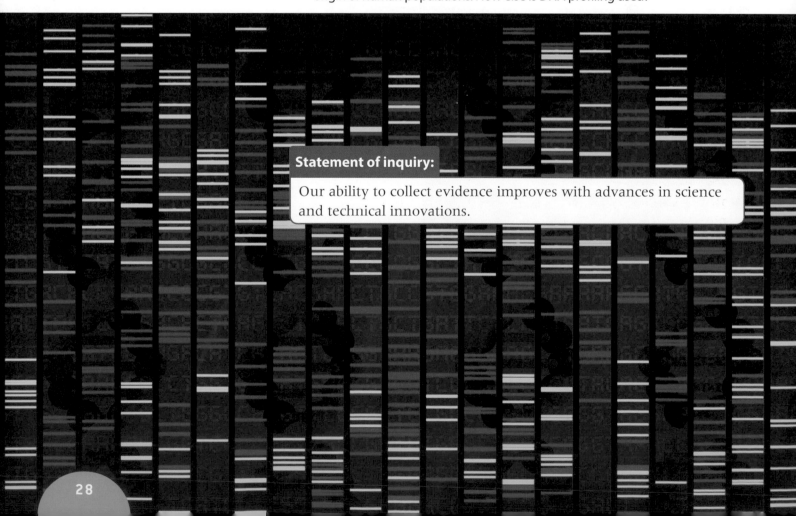

Statement of inquiry:

Our ability to collect evidence improves with advances in science and technical innovations.

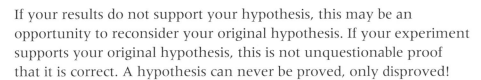

If your results do not support your hypothesis, this may be an opportunity to reconsider your original hypothesis. If your experiment supports your original hypothesis, this is not unquestionable proof that it is correct. A hypothesis can never be proved, only disproved!

Whether you are considering qualitative or quantitative evidence, you are trying to identify relationships between variables. For this reason, the key concept of this chapter is relationships.

Scientists use curiosity, creativity, imagination and intuition to design and undertake investigations. Advances in our understanding of the chemical world in the fields of medicine, agriculture, engineering, chemical industries and synthetic chemistry are some of the areas of development that have helped to shape our modern world. These endeavors use evidence obtained from an advancing level of scientific technology to promote our understanding. For this reason, the global context of this chapter is scientific and technical innovation.

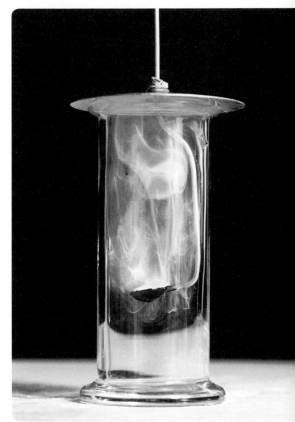

▲ There is a well-established path for the development of new pharmaceutical drugs. It starts with research in the laboratory, followed by preclinical research to find out more about the safety of the drug. Clinical research involving trials on humans can only start if the results of this phase are satisfactory. The clinical research phase of the development cycle usually lasts many years and involves thousands of volunteers in drug trials. In the US, the Food and Drug Administration (FDA) review all the data and evidence provided by the pharmaceutical companies and decide whether or not to approve it. At every stage of this process, evidence is at the centre

▲ Lithium, sodium and potassium are in the same group in the periodic table. Potassium reacts more violently with water than sodium, which in turn reacts more violently than lithium. The photo shows what happens when sodium reacts with oxygen. What is your hypothesis about the violence of the reaction of lithium with oxygen?

▲ The addition of these two colorless solutions, lead(II) nitrate and potassium iodide, results in the formation of a bright yellow precipitate of lead(II) iodide. Can observations alone inform us that a reaction has occurred?

What do you observe?

In our daily lives, we use the senses of sight, smell, touch, hearing, and taste to interact with the environment. Most of the time we do this subconsciously but when we are performing experiments our awareness is heightened. What senses do you use to make observations during an experiment?

In an investigation, there are two types of observation that you can make.

Qualitative observations

Macroscopic properties are those you can observe with your eyes, unaided by technology. These are called qualitative observations. For example, when you add two chemicals together you may observe a color change, the formation of a precipitate or solid, the release of gas, or solids dissolving or reacting with solvents.

Be careful not to confuse an inference with an observation. For example, if you add powdered zinc to an acid solution, the mixture bubbles—this can be observed. You may believe that hydrogen gas is being produced, but this is an inference. An inference is a prediction. You cannot identify hydrogen gas by looking at the bubbles; you have to perform an analytical test to determine its composition. Observations help us to establish the relationships that exist between reactants and products.

Quantitative observations

These involve the collection of numerical data during an experiment. Instrumentation is required, such as an electronic balance, thermometer or gas syringe. The data collected is then analyzed and a conclusion based on the data is drawn.

▶ A pH meter is being used to record the pH of a solution. Is examining the changing color of litmus paper a qualitative or quantitative observation?

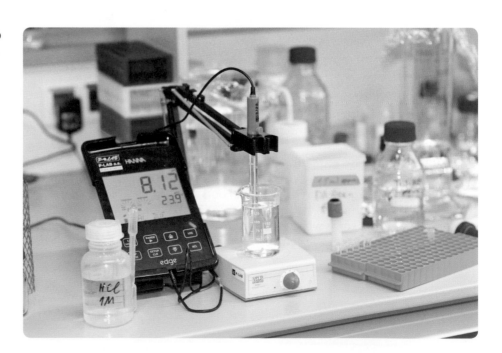

ATL Critical thinking skills

Using inductive and deductive reasoning

Deductive reasoning is what you recognise as the scientific method. Having formulated a testable hypothesis, deductive inferences are made. This is where we make predictions about the observations that we will make based on our understanding of the theory, that we hold to be true. With deductive reasoning we are moving from the general, in the form of the theory, to the specific in the form of our observations. For example, we learnt that the Group 18 noble gases are generally unreactive because they have full outer electron shells; neon is a noble gas, so deductive reasoning tells us that neon is unreactive.

▲ Deductive reasoning informs that if we can only pick cherries from this tree, this tree must be a cherry tree

Inductive reasoning is the opposite of deductive reasoning, as our thinking moves from the specific to the general. With inductive reasoning, we consider the evidence collected from experimental observations and make broad general statements based on it. The conclusions drawn from this evidence can be used to make sense of the theories we are investigating. For example, upon dropping a piece of sodium metal into water, we observe a violent reaction; sodium is an alkali meta, so using inductive reasoning, we could conclude that all alkali metals react with water.

1. What roles do deductive and inductive reasoning play in helping you make sense of your everyday experiences?

2. How do deductive and inductive reasoning help you with your study of science?

How was the periodic table developed?

The modern periodic table is split into groups (columns) and periods (rows). The elements are arranged in order of increasing atomic number. The arrangement of elements in the periodic table enables chemists to predict the physical properties and chemical reactions from the elements involved.

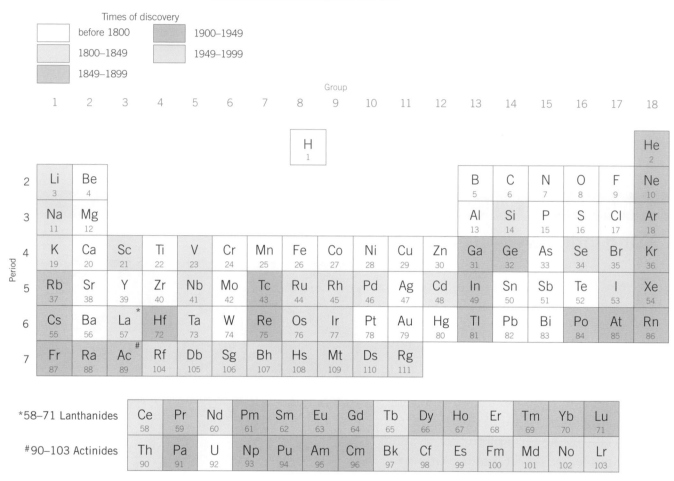

In the modern periodic table elements are arranged in order of increasing atomic number. It is based on the 1913 work of British physicist Henry Moseley

Online periodic tables

There are many online versions of the periodic table providing information about each of the elements. Search online for two or three versions.

Now choose an element and compare the information each periodic table gives about it. Which version do you think is best and why? Compare your conclusions with the rest of the class.

ATL Critical thinking skills

Understanding based on new information and evidence

Evidence is used to develop theories. These theories enable us to make predictions which can then be challenged to see if they are valid. The periodic table is a good example of this.

Research how the ideas of Dobereiner and Newlands and the groundbreaking proposals of Mendeleev and Moseley led to what is known today as the modern periodic table. Consider the following.

- **Johann Dobereiner's law of triads:** Although this did not include many elements, it suggested an important link between atomic weight and what other factor? Why was it an important step forward in our understanding?

- **John Newlands' law of octaves:** This proposed that elements displayed periodicity or repeating patterns. What was the significance of this?

- **Dmitri Mendeleev's periodic table:** Mendeleev left gaps for elements that had not yet been discovered. He predicted their properties and placed them in his periodic table where he thought they should be. How successful were his predictions?

- **Henry Moseley's modern periodic table:** How did Moseley's approach differ from that of Mendeleev?

What characterizes a metal?

GENERAL

The periodic table is often separated into metallic elements and non-metallic elements, but what is a metal? Characteristically, metals:

- are good conductors of heat and electricity
- are malleable—they can be hammered into different shapes
- are ductile—they can be drawn into wires
- are shiny when pure and unreacted
- are sonorous—they ring when struck
- tend to have high densities
- have high melting and boiling points
- tend to lose electrons to form cations (positively charged ions).

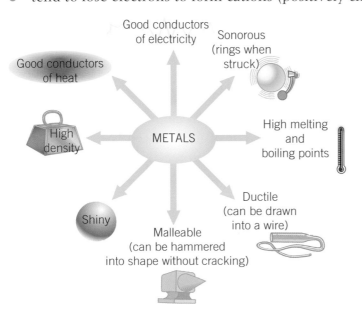

▲ Characteristic properties of metals

The elements of group 1, the alkali metals, demonstrate predictable chemical and physical properties. They are increasingly reactive as you move down the group; this is demonstrated in their reactions with water. Chapter 11, Patterns, considers this further.

 Experiment

Reaction of alkali metals with water

During this demonstration of the reactions of alkali metals and water, carefully observe the reactions, and record your observations.

⚠ Safety

Students should remain a safe distance from the water trough. The use of a transparent screen in front of the tank is highly recommended.

Materials

- Thick-walled glass water trough
- Transparent screen
- Universal indicator (UI) solution (see Chapter 3, Consequences)
- Sodium and potassium metal stored in oil
- Wooden splint, filter paper, tweezers and knife

Method

1. Add a few drops of UI solution to the water-filled trough.

2. Using tweezers, remove a small piece of sodium from the oil and place it on a piece of filter paper. Using a knife, cut a small piece of sodium metal and transfer it into the water using tweezers.

3. Wait until the initial piece of metal has fully reacted with the water before adding any further pieces.

4. Light a wooden splint and place the flame close to the reacting metal. (Look at Chapter 3, Consequences, to remind yourself about this test.)

5. Change the water in the trough and repeat using the potassium.

Questions

1. List three observations of what happens when sodium and potassium metal are dropped into water.

2. What flame color did you observe from each reacting metal?

3. What did you notice about the level of reactivity of sodium compared to potassium?

4. What does the color of the UI tell you about the initial and final pH of the water solution? Estimate the pH of the resulting solution.

5. Write word equations to describe the reactions of sodium and potassium with water.

6. Using the skills developed in Chapter 1, Balance, construct balanced chemical equations for both reactions.

7. Consider some typical metals and their physical and chemical characteristics. For example, a lead fishing sinker is denser than water and sinks to the bottom of the lake; an iron nail corrodes in air, but very slowly. Consider what happens if you drop a piece of copper, aluminium or iron into water. Describe how the behavior of the alkali metals are different from other metals.

▲ When sodium is added to water, it reacts violently

What characterizes non-metals?

GENERAL

Non-metal elements tend to have properties that are the opposite to metals. Characteristically, non-metals:

- are poor conductors of both heat and electricity
- are brittle
- tend to have low densities
- have low melting points and boiling points
- are non-ductile
- tend to gain electrons and form anions (negatively charged ions).

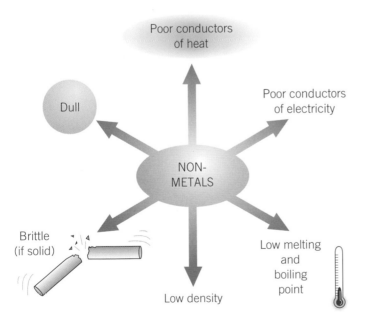

Poor conductors of heat

Dull

Poor conductors of electricity

NON-METALS

Brittle (if solid)

Low density

Low melting and boiling point

▲ The characteristics of non-metals are very different to those of metals

What are transition elements?

The transition elements are found in groups 3 to 12 of the periodic table. While the properties of elements in groups such as the alkali metals (group 1) and the halogens (group 17) follow patterns that can be predicted, the properties of transition elements are less predictable. However they have some common characteristics including:

- they are all metals, which is why they are often called transition metals

- they form colorful compounds

- they can be used as catalysts—a catalyst speeds up the rate of a reaction, but is not consumed in the reaction

- many can form ions with different charges, for example, chromium can form Cr^{3+} and Cr^{6+} ions (as well as others).

▲ Transition metal salts are formed by the reaction between a transition metal and an acid. These colorful metal salts include, on the front row, from left to right, cobalt(II) chloride, copper(II) carbonate, iron(II) sulfate and copper(II) sulfate; and on the back row, manganese(II) carbonate, nickel(II) sulfate, potassium chromate(VI) and chromium(III) potassium sulfate

Changing the oxidation states of vanadium

Vanadium is a silvery metal that resists corrosion. It is added to steel to increase its strength. It exists in many different compound as it has multiple oxidation states. It can form V^{2+}, V^{3+}, V^{4+} and V^{5+} ions, and compounds containing these ions are different colors.

Find a good demonstration of the changing oxidation states of vanadium on the internet. (We are not suggesting you work with these chemicals yourself as the chemicals used are both toxic and corrosive, and the waste generated from the experiment is difficult to dispose of safely.)

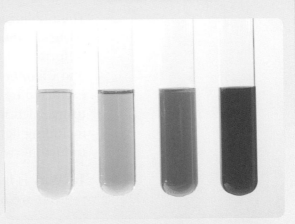

▲ Solutions of dissolved vanadium compounds in different oxidation states

1. Describe your observations.

2. What evidence did you observe to suggest chemical reactions were occurring?

3. How many different oxidation states of vanadium were formed?

4. List the oxidation states and their colors.

5. Suggest how the characteristic colors of transition metal compounds can be helpful when performing investigations.

Harvesting vanadium

The green-barrel sea squirt is native to the Indo-Pacific Ocean region. It lives attached to rocks and dead coral. A filter feeder, it draws in and expels seawater to extract required nutrients such as plankton and other forms of matter. Sea squirts accumulate heavy metals such as vanadium from seawater. They contain vanabins, a vanadium-binding metalloprotein in their blood cells. In the same way that the hemoglobin molecule in human red blood cells binds to oxygen molecules, vanabins loosely bind to vanadium atoms before releasing them into the bloodstream of the sea squirt. At present, it is not known why sea squirts extract vanadium from seawater. The blood of sea squirts can contain 100 times the concentration of vanadium present in seawater.

1 Vanadium is added to steel and is increasingly used in the production of batteries. Can an increasing demand for vanadium justify harvesting sea squirts for vanadium? Debate the pros and cons of this proposal.

What are noble gases?

The noble gases are found in group 18 of the periodic table. They are naturally **inert** (unreactive) giving them stability and they tend not to form ions naturally (we look at this further in Chapter 9, Models and Chapter 11, Patterns). Noble gases are present in air and can be extracted by liquefying air and then separating out the noble gases using fractional distillation. They have a wide range of applications including lighting, refrigeration, air ships and balloons, as well as in industrial welding.

1. Why is it an advantage for chemists to have access to elements that are inert?

2. Describe some applications in which this property can be exploited.

▶ Liquid helium has a very low boiling point of 4 K (−269°C). It is used to cool the superconducting magnets in magnetic resonance imaging (MRI) scanners

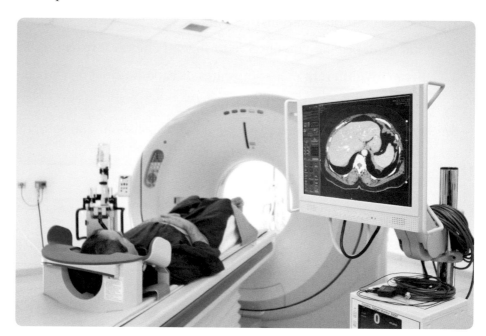

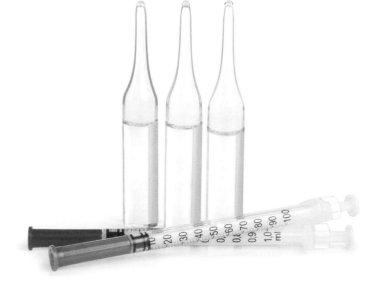

▶ Ampoules can be used to isolate vaccines from atmospheric oxygen and other contaminants. They are often filled with inert gases such as nitrogen or noble gases such as argon

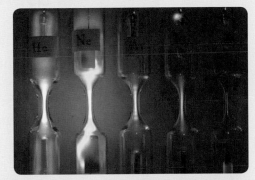

What are fossil fuels?

ORGANIC

Organic chemistry is the chemistry of carbon-containing compounds. Organic compounds are one of the most important groups of chemicals. They include fuels, paints, dyes, alcohol, plastics, industrial solvents, drugs and medicines, foods, pesticides and fertilizers.

Globally, there are three main fossil fuels upon which we have become dependent: oil, coal and natural gas. Fossil fuels are the product of the decomposition of carbon-based life forms such as marine animals and plants that captured the Sun's energy millions of years ago. They take millions of years to develop and therefore are classified as a non-renewable energy source.

As economies have developed over the last two centuries, our dependence on non-renewable fossil fuels has grown. In 2016, more than 35 countries had greater than 90% dependency on fossil fuels for their energy needs. In countries such as India and China, heavy industry has grown and lifestyles have changed; there is also a vast increase in the number of cars being manufactured and driven on their roads. Daily global consumption is close to 100 million barrels of oil a day.

1 Why does Malta, which is 99% dependent on fossil fuels, have less of an impact on global demand than China which is 88% dependent on fossil fuels?

▲ Offshore oil reserves are sometimes the subject of territorial disputes between nations. The demand for fossil fuels is so great, countries strongly defend their oil reserves

ORGANIC

How is crude oil transformed into useful fuels?

Once crude oil has been extracted, it needs to be refined before the hydrocarbons present can be fully utilized. Hydrocarbons are compounds that contain carbon and hydrogen atoms only. Butane, C_4H_{10}, is a flammable hydrocarbon used in portable stoves as a fuel source. Glucose, $C_6H_{12}O_6$, contains oxygen as well as carbon and hydrogen; it is an example of a carbohydrate, not a hydrocarbon.

Crude oil is a mixture of short- and long-chain hydrocarbon molecules and varying amounts of organic compounds containing nitrogen, oxygen and sulfur. This is discussed in greater detail in Chapter 6, Form. Not all crude oil is the same. Analysis of samples of crude oil provides evidence of its origin. Typically, sulfur content varies from one location to the next; this also affects the cost of refining. In 2017, the major crude oil-producing nations were Russia, Saudi Arabia, the USA and China.

Fractional distillation is the industrial process used to separate crude oil into its useful components. The crude oil is heated causing small-chain, highly volatile hydrocarbons with low boiling points to rise to the top of the fractionating column. The longer the hydrocarbon chain, the higher the boiling point, and the less they rise up the column. The gases produced are condensed and collected for further purification.

Mid-length hydrocarbons were often considered to be a residue of the process used to make fuels for transportation, but now they are broken down into smaller molecules of alkanes and alkenes which are invaluable to the chemical industry.

▶ You can often find fractional distillation towers around port facilities of large oil terminals. What are the advantages of locating an oil refinery close to the supply network for the crude oil?

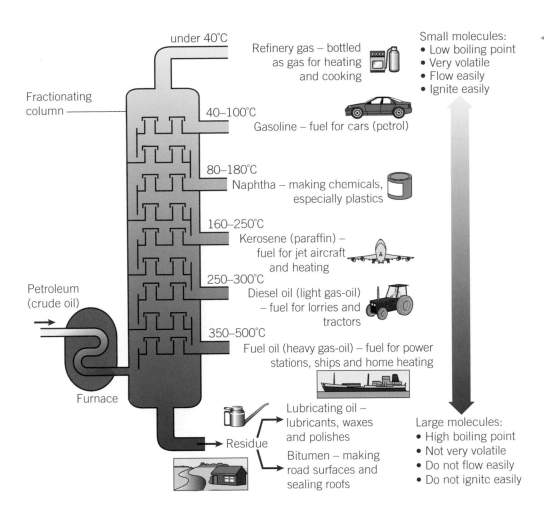

under 40°C

Refinery gas – bottled as gas for heating and cooking

Small molecules:
• Low boiling point
• Very volatile
• Flow easily
• Ignite easily

Fractionating column

40–100°C

Gasoline – fuel for cars (petrol)

80–180°C

Naphtha – making chemicals, especially plastics

160–250°C

Kerosene (paraffin) – fuel for jet aircraft and heating

250–300°C

Petroleum (crude oil)

Diesel oil (light gas-oil) – fuel for lorries and tractors

350–500°C

Fuel oil (heavy gas-oil) – fuel for power stations, ships and home heating

Furnace

Lubricating oil – lubricants, waxes and polishes

Large molecules:
• High boiling point
• Not very volatile
• Do not flow easily
• Do not ignite easily

Residue

Bitumen – making road surfaces and sealing roofs

◄ Fractional distillation is a relatively simple physical separation technique that utilizes differences in the boiling points of hydrocarbon fractions. It has enabled large-scale refining of crude oil to meet the global demand for the various fractions such as gasoline for road transport and kerosene for the aviation industry. Chemists working in the petrochemical industry analyze evidence about the composition of the fractions and then adjust the parameters and conditions of this industrial process to maximize yield and profits

Data-based question: How does the composition of crude oil reserves vary?

Not all crude oil is the same. The composition varies according to the oil reserve from which it is extracted.

1. Look at the charts and the illustration of fractional distillation. Explain which crude oil is best suited to the production of gasoline.

2. Explain why crude oil from Venezuela is called heavy oil. How does the molecular mass of the most common hydrocarbon in this crude oil differ from the hydrocarbons commonly found in Saharan crude oil?

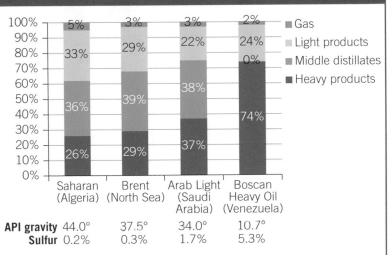

	Saharan (Algeria)	Brent (North Sea)	Arab Light (Saudi Arabia)	Boscan Heavy Oil (Venezuela)
Gas	5%	3%	3%	2%
Light products	33%	29%	22%	24%
Middle distillates	36%	39%	38%	0%
Heavy products	26%	29%	37%	74%
API gravity	44.0°	37.5°	34.0°	10.7°
Sulfur	0.2%	0.3%	1.7%	5.3%

▲ Refinery output for different crude oils. API stands for the American Petroleum Institute. API gravity is a measure of the density of the crude oil. A higher API gravity indicates less dense oil

3. The terms sweet crude oil and sour crude oil refer to the sulfur content of the crude oil. Crude oil with less than 0.5% sulfur content is classified as sweet. Which crudes on this graph are considered sweet oil?

4. Sulfur is corrosive. Suggest reasons why sweet oil is easier to refine and is preferred by refineries? Why does sweet oil command a higher price per barrel?

ORGANIC

Why are naming rules for chemical compounds important?

The International Union of Pure and Applied Chemistry (IUPAC) was founded in 1919. This followed the recognition that as research became more international and multilingual, a common system of weights, measures, nomenclature and symbols was needed to facilitate the rapid and clear communication of new ideas. Once the evidence is established for the composition of a new compound, IUPAC devises the official name of the compound.

▶ Taxol is an anticancer drug that is used post-surgery in combination with chemotherapy to prevent the recurrence of breast cancer. The IUPAC name accurately describing the structure of this molecule is (2α,4α,5β,7β,10β,13α)-4,10-bis(acetyloxy)-13-{[(2R,3S)- 3-(benzoylamino)-2-hydroxy-3-phenylpropanoyl]oxy}-1,7-dihydroxy-9-oxo-5,20-epoxytax-11-en-2-yl benzoate

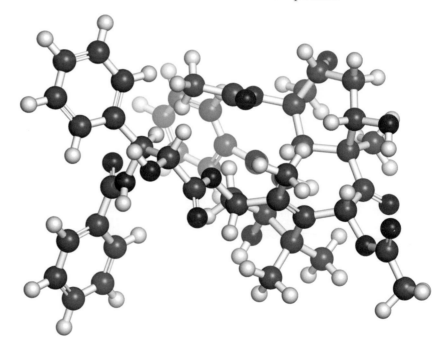

ORGANIC

What are alkanes and where do they come from?

Alkanes are a family of saturated hydrocarbons that only contain hydrogen and carbon atoms. Saturated means that they only contain carbon–carbon single covalent bonds (see Chapter 9, Models). They are found in unprocessed crude oil. After being refined from crude oil, molecules such as butane and octane are used in combustion reactions as a fuel, releasing energy. The combustion of butane is described by the following equation:

$$2C_4H_{10}(g) + 13O_2(g) \longrightarrow 8CO_2(g) + 10H_2O(g)$$

$$\Delta H^0 = -2877 \text{ kJ mol}^{-1}$$

IUPAC nomenclature for alkanes

You have probably heard the names of some alkanes before: methane, propane, butane and octane are all types of fuels. The rules for the naming of the alkane series form the foundation of the naming of all organic compounds. Understanding these rules is important to understanding organic chemistry.

The structure of an organic compound can be represented in a number of different ways.

- The molecular formula describes the actual number of atoms found in a molecule of the compound. For example, C_3H_8 is the molecular formula of the alkane, propane.

- The full structural formula of a compound is a two-dimensional representation of all atoms, the bonds and their relative positions.

◀ IUPAC root names for the alkane series

Number of carbons	Root name	Molecular formula	Structural formula
1	methane	CH_4	H | H — C — H | H
2	ethane	C_2H_6	H H | | H — C — C — H | | H H
3	propane	C_3H_8	H H H | | | H — C — C — C — H | | | H H H
4	butane	C_4H_{10}	H H H H | | | | H — C — C — C — C — H | | | | H H H H
5	pentane	C_5H_{12}	H H H H H | | | | | H — C — C — C — C — C — H | | | | | H H H H H
6	hexane	C_6H_{14}	
7	heptane	C_7H_{16}	
8	octane	C_8H_{18}	
9	nonane	C_9H_{20}	
10	decane	$C_{10}H_{22}$	

1. Determine the relationship between the number of carbons and the hydrogens in alkanes: if an alkane has n carbons, how many hydrogens does it have? Write your answer in the form $C_nH_?$.

2. Where else do we use rules to govern communication in both writing and speech? Why are these protocols essential for clear communication of meaning?

Naming alkanes with side chains

Sometimes an alkane has one or more of its hydrogens replaced with a side chain or substituent. Carefully read these rules which show you how to work out the name of a substituted alkane.

1. Examine the structure of the compound. Identify the longest continuous carbon chain. This is called the parent chain and provides the root name.

   For this compound, the parent chain has six carbons so the root name is **hexane**.

2. Identify side chains or substituents. The names of the alkyl substituents used in the IUPAC nomenclature system are listed in the table below. If there are more substituents than one, list them in alphabetical order.

   Here there is a $-CH_3$ side chain; this a **methyl** group.

Substituent name	Condensed formula	Structural formula
methyl	$-CH_3$	
ethyl	$-CH_2CH_3$	
propyl	$-CH_2CH_2CH_3$	
butyl	$-CH_2CH_2CH_2CH_3$	

3. Number the longest chain. The direction of the numbering depends on the position of any substituents. The substituent must be positioned on the lowest numbered carbon atom.

   ✓   ✗

   Here, the correct position of the methyl group is carbon-**2**.

4. Write the full name of the compound.

   In this example, the compound is 2-methylhexane.

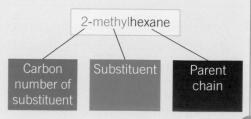

2-methylhexane

Carbon number of substituent

Substituent

Parent chain

5. When two or more substituents are positioned in equivalent positions on the parent chain, number the chain so that the substituent which comes first alphabetically is positioned on the lowest numbered carbon atom. Look at the example of 3-ethyl-7-methylnonane: the **ethyl** group has priority over the **methyl** group so the carbon chain is numbered to give the lowest possible numbered carbon for the ethyl group, in this case carbon-3.

methyl group

ethyl group

6. Multiple substituents of the same type are indicated by using numerical multipliers as shown in the table below. For example, "di" is used for two identical substituents.

Number of substituents	Prefix
1	(mono)
2	di
3	tri
4	tetra

▲ Numerical multipliers in the IUPAC nomenclature system

In 2,4-dimethylpentane "di" indicates that there are two identical substituents.

Note the following details about how we write this name:

● a comma is used to separate numbers within the name

● a hyphen is used to separate numbers and letters.

1. Draw the structural formula for this hydrocarbon after analyzing the structure and its IUPAC name.

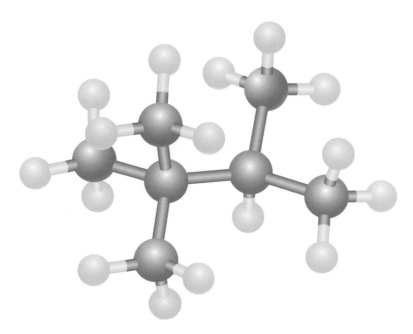

◄ The compound 2,2,3-trimethylbutane is an organic compound used as a fuel additive to improve the combustion of fuel in car engines

ORGANIC

How do alkenes differ from alkanes?

Alkenes are an important group of organic compounds. They differ from alkanes because they contain carbon-carbon double bonds, and because of this contain fewer hydrogens.

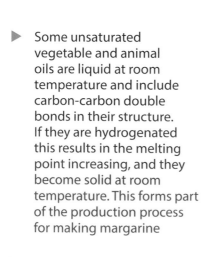

▲ In butane (an alkane), each carbon atom forms four bonds with other atoms

▲ Here the carbon atoms involved in the double bond can only form two further bonds with other atoms

- The formula of ethane, an alkane, is C_2H_6. The general formula for alkanes is C_nH_{2n+2}.

- The formula for ethene, a member of the alkene class, is C_2H_4. The general formula for alkenes is C_nH_{2n}.

Carbon is found in group 14 of the periodic table. It has four valence electrons and can form four bonds with other atoms.

Sometimes a carbon atom will form a double bond with another carbon atom. This means that it can only form two further bonds.

We describe compounds like alkanes, without any carbon-carbon double bonds, as saturated. Alkenes are described as being unsaturated, as the carbon-carbon double bond can be broken, leaving the affected carbon atoms in a position to form bonds with other atoms. Alkenes are therefore more reactive than alkanes.

Alkenes are commonly involved in synthetic chemical reactions. A synthesis reaction is the formation of a single product from two or more simpler reactants. For example, ethene reacts with hydrogen to form ethane in a process that is called hydrogenation:

$$C_2H_4(g) + H_2(g) \xrightarrow[\text{catalyst}]{\text{heat, pressure}} C_2H_6(g)$$
$$\text{ethene} \qquad\qquad\qquad\qquad\qquad \text{ethane}$$

1. Re-write the equation for the hydrogenation of ethene using structural formulae.

▶ Some unsaturated vegetable and animal oils are liquid at room temperature and include carbon-carbon double bonds in their structure. If they are hydrogenated this results in the melting point increasing, and they become solid at room temperature. This forms part of the production process for making margarine

Identifying alkenes

If bromine water is mixed with an alkene (for example, ethene) the bromine water goes from an orange–brown color to colorless. This reaction is used to indicate the presence of a carbon-carbon double bond.

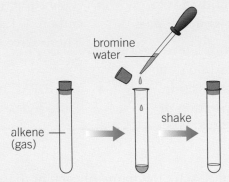

1. What reaction could be happening between bromine water and the alkene to cause the color change?

The rules for naming alkenes are similar to those for alkanes. Work through the steps below to learn how to name them.

1. Analyze the structure of the compound to find the longest continuous carbon chain. This is called the parent chain and provides the root name.

2. The suffix to indicate the presence of at least one carbon-carbon double bond changes from -**ane** to -**ene**.

3. Number the carbons in the longest chain so that the carbon-carbon double bond is at the lowest possible number. This takes priority over the position of any substituents.

Worked example: Naming alkenes

Question

Name the following alkenes.

(a)

$$H - \underset{\underset{H}{|}}{\overset{\overset{H}{|}}{C}}_4 - \underset{\underset{H}{|}}{\overset{\overset{H}{|}}{C}}_3 - \underset{\underset{H}{|}}{\overset{\overset{}{}}{C}}_2 = C_1 \overset{H}{\underset{H}{}}$$

(b)

$$H - \underset{\underset{H}{|}}{\overset{\overset{H}{|}}{C}}_1 - \overset{\overset{H}{|}}{C}_2 = \overset{\overset{H}{|}}{C}_3 - \underset{|}{\overset{\overset{H}{|}}{C}}_4 - \underset{\underset{H}{|}}{\overset{\overset{H}{|}}{C}}_5 - \underset{\underset{H}{|}}{\overset{\overset{H}{|}}{C}}_6 - H$$
$$H - \underset{\underset{H}{|}}{\overset{}{C}} - H$$

Answer for structure (a)

- The longest carbon chain is four carbons long so the parent chain is **but**-.

- The molecule contains a carbon-carbon double bond so the suffix is -**ene** not -ane.

- Number the carbon chain from right to left so the unsaturated carbon-carbon double bond is in the lowest possible position. The double bond follows carbon-**1**.

- The molecule is called 1-butene.

Answer for structure (b)

- The longest carbon chain is six carbons long, and so the parent chain is **hex-**.

- The molecule contains a carbon-carbon double bond so the suffix is **-ene** not **-ane**.

- Number the carbon chain from left to right so the unsaturated carbon-carbon double bond is in the lowest possible position. This takes priority over the position of the substituent. The double bond follows carbon-2.

- A substituent/side chain comes off carbon-4. It is one carbon in length and is a derivative of the methane, so is a **methyl** group.

- The IUPAC name for this molecule is 4-methyl-2-hexene.

1. Apply IUPAC nomenclature rules, state the name of each of these molecules.

a)

b)

c)

d)

e)

Experiment

Catalytic cracking of paraffin

Catalytic cracking is the chemical procedure used to break up large hydrocarbon molecules into a mixture of smaller alkanes and alkenes. Naphtha, a fraction from crude oil distillation, is often the source of these large hydrocarbons. These smaller-chain alkanes can be used as a fuel and smaller alkenes, such as ethene, are used to make polymers in the plastics industry.

Paraffin is another fraction that is derived from crude oil. It is a mixture of alkanes and alkenes, and it can be cracked in a school laboratory.

⚠ Safety

- This experiment should only be carefully performed by students who are familiar with laboratory procedures. Alternatively, this can be a teacher demonstration.

- Wear safety glasses.

- Use a safety screen and ensure the room is well ventilated.

- The bromine water and potassium permanganate(VII) solution must be be correctly disposed of at the end of the demonstration.

- Be aware that suck-back occurs when the gas inside the boiling tube cools and condenses rapidly, reducing the pressure inside it. This causes the water from the trough to be sucked up the discharge tube and into the hot boiling tube. This may result in the glass cracking.

Materials

- Liquid paraffin

- Catalyst: crushed pumice stone or porous pot fragments

- 0.01 mol dm^{-3} bromine water

- 0.002 mol dm^{-3} acidified potassium permanganate(VII) solution

- Mineral wool

- Boiling tube

- Dropping pipette

- Water trough

- Bunsen burner

- Stand and clamps

- Six test tubes

- Rubber bungs, glass tubing, spatulas, and wooden splints

Method

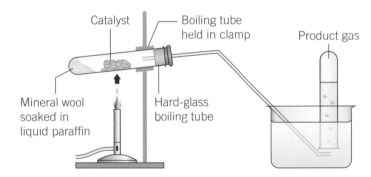

1. Put mineral wool to a depth of 2 cm in the bottom of the boiling tube. Drop 2 cm^3 of liquid paraffin onto it using a dropping pipette.

2. Clamp the boiling tube as shown in the diagram.

3. Place 2–3 spatulas of the catalyst in the centre of the tube and connect the tube.

4. Fill six test tubes with water and stand them inverted in the water trough.

5. Strongly heat the catalyst in the middle of the tube for a few minutes. Do not heat the boiling tube too close to the rubber bung.

6. When the catalyst is hot, pass the flame back and forth over the liquid paraffin and catalyst to vaporize some of the liquid paraffin.

7. Once the gas starts to be released, it is essential to continue heating the contents of the boiling tube to avoid suck-back. If it looks like suck-back is about to occur, lift the gas discharge tube from the water.

8. With a steady stream of gas bubbles established, collect six test tubes of gas, sealing each full test tube with a bung. Keep the first two test tubes separate from the others.

9. When gas collection is complete, first remove the delivery tube from the water by tilting or lifting the clamp stand. Only then stop heating.

Now you will need to perform some analytical tests on the gas collected.

Analytical tests procedure

1. Discard the first two test tubes of gas collected.

2. Take the third test tube, remove the bung and gently waft the gas towards you. What does the smell remind you of?

3. Is the gas flammable? Take the fourth test tube, remove the bung and place a lit splint into the mouth of the test tube. What happens?

4. To the fifth test tube, add 2–3 drops of bromine water, stopper it and shake well. What do you observe?

5. To the sixth test tube, add 2–3 drops of acidified potassium permanganate(VII) solution, stopper and shake well. What do you observe?

Questions

1. Explain why the first two test tubes are discarded. What is the main component in these samples?

2. Describe the color changes observed in the tests performed on the final two test tubes.

3. Predict the outcome of both of these tests on the first two test tubes that were discarded. Explain your reasoning.

ORGANIC

What are alcohols and their applications?

Alcohols are a class of organic compound that contain the functional group –OH. A functional group is the reactive part of an organic molecule and often contains oxygen or nitrogen.

In alcohols, the hydroxyl group –OH replaces one hydrogen of a hydrocarbon molecule. Alcohols have a wide range of applications in industry and play an important role in synthesis reactions.

Alcohols can be made in large volumes from alkenes. For example, ethanol is produced in the reaction between ethene and steam in the presence of a catalyst (solid silicon dioxide coated in phosphoric(V) acid). Ethanol is used in alcoholic drinks, as an alternate fuel to gasoline and as a solvent for many reactions.

▲ Alcohols contain the functional hydroxyl group

Ethene $C_2H_4(g)$

Water $H_2O(g)$

300°C, 60 atm
Phosphoric(V) acid catalyst

Ethanol $C_2H_5OH(g)$

Worked example: Naming alcohols

Question

When naming alcohols, the suffix of the principal group is positioned after the parent chain. What is the name of this compound?

$$H-\underset{\underset{H}{|}}{\overset{\overset{H}{|}}{C}}-\underset{\underset{H}{|}}{\overset{\overset{H}{|}}{C}}-OH$$

Answer

- The compound has two carbon atoms so the parent chain is eth-.
- There are only single carbon–carbon bonds so the suffix after the parent chain is -an(e).
- The functional group is the hydroxyl group so the suffix is -ol.
- The IUPAC name for this organic compound is **ethanol**.

Propan-2-ol   2-methylpropan-1-ol   ◀ Examples of other alcohols

Data-based question: Ethanol and biofuels

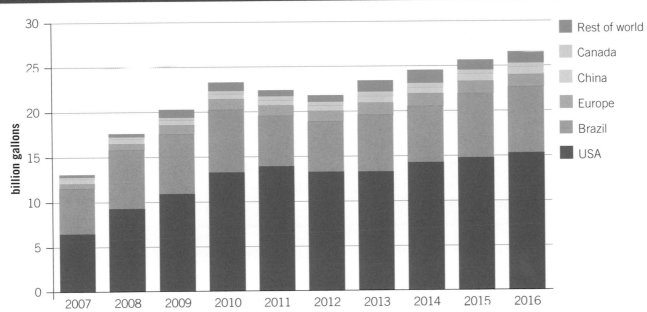

▲ Global ethanol production from 2007–2016. Biofuels such as ethanol and biodiesel are a renewable energy source. The hydrocarbons in these fuels come from organic matter such as sugar cane, which can be grown in a matter of months. In contrast, fossil fuels take millions of years to form. This is the reason why fossil fuels are considered to be non-renewable

The text below is part of an online article from the US Department of Energy from 13 October 2016. Note the following details before you read the article.

- E85 is a blend of gasoline and ethanol which contains up to 85% ethanol.

- A US gallon is an imperial unit of measurement of volume. It is the equivalent of 3.79 litres.

- A US ton is an imperial unit of measurement of mass. It is the equivalent of 907 kg.

California ramps up biofuels infrastructure

New renewable diesel, biodiesel, and E85 fueling pumps are cropping up across California as Clean Cities stakeholders work to expand the availability of biofuels.

In 1996, the Golden State had only one E85 station. As of 2016, it boasts more than 95 public stations, due in large part to a station-development project known as the Low Carbon Fuel Infrastructure Investment Initiative (LCF13). The project has the potential to displace 39 million gallons of petroleum and 187,500 tons of carbon dioxide emissions per year.

In addition, the project has created more than 450 green jobs in a state particularly hard-hit by unemployment. LCF13 is funded through an American Recovery and Reinvestment Act award and by the California Energy Commission and Propel Fuels. California has more than 1 million diesel and flexible fuel vehicles on the road, and LCF13 is now matching the fuels to the vehicles.

East Bay Clean Cities and other California coalitions are working with project partners, like Propel Fuels and the California Department of General Service, to target ZIP codes with high densities of alternative fuel vehicles and to find locations where the coalitions' fleet partners operate. Clean Cities coalitions are also coordinating with station owners to publicize the new facilities, through station-opening events, ethanol buy-down days, and identification and outreach to potential fleet customers for the new fueling facilities.

Biofuels providers and Clean Cities coalitions began laying the foundation for the new fueling infrastructure back in 2007 by contacting elected officials, educating local regulatory agencies, and helping to streamline permitting processes. Today, fuel providers are able to offer station owners turn-key packages, complete with tanks and dispensers, with a minimum of paperwork. And with more stations coming online every month, California is poised to be a global leader in the deployment of low-carbon, alternative fuels.

1. Identify the largest producer of ethanol and suggest possible reasons for this.

2. Discuss why the level of production of ethanol has increased from 2007 to 2016.

3. China has a large population in global terms and a very large manufacturing sector, but its ethanol production is among the lowest of the nations listed. Why might that be the case? Do you think that this might change in the near future, and if so why?

4. What natural resources might explain Brazil's high ethanol production?

5. Summarize this article for an international audience in no more than 150 words.

6. What benefits has LCF13 brought to California?

7. In a small group, discuss any biofuel use and initiatives locally, or in your country. Do you think the use of biofuels in your country is important for the future? Explain why or why not?

Representing organic molecules

In organic chemistry, molecular formulae are not used often as they do not give any useful information about the bonding in the molecule. Structural formulae tell you about the bonding, but they can be drawn in different ways.

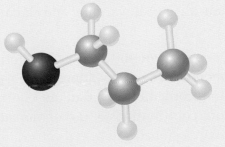

$CH_3{-}CH{=}CH{-}CH{-}CH_2{-}CH_3$
CH_3

◀ Two ways of representing 4-methyl-2-hexene

Sometimes it is helpful to look at 3D models of organic molecules.

Match these models (1–3) with their structural formulae (a–c) and work out their IUPAC names.

1.

2.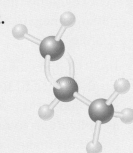

3.

a. $CH_3{-}CH{=}CH_2$ b. $CH_3{-}CH_2{-}CH_3$ c. $CH_3{-}CH_2{-}CH_2{-}OH$

What is the composition of the Earth's atmosphere?

ENVIRONMENTAL

Atmospheric air is a mixture of gases and microscopic particulate matter. The largest component by volume is nitrogen (78%). Oxygen makes up approximately 21% and carbon dioxide about 400 ppm (this means that for every million air particles, 400 of them are carbon dioxide) or 0.04%. The rest is mainly argon and other noble gases, such as helium, neon, krypton and xenon.

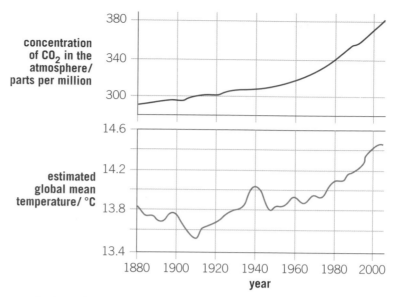

▲ Carbon dioxide and global mean temperatures

Average temperatures around the world are increasing—this is called global warming. Most scientists agree that the main cause of this is the increase in the levels of greenhouse gases, in particular, carbon dioxide and, to a lesser extent, methane.

1 What evidence do we have that the combustion of fossil fuels is having a negative impact, causing global warming and climate change?

2 In 2015, the global benchmark value of 400 parts per million for the concentration of carbon dioxide in the atmosphere was exceeded for the first time in recorded history. Can our efforts as individuals to minimize our contributions of carbon dioxide emissions make a difference on a global scale?

ATL Information and media literacy skills

Communicate information and ideas effectively

At the Paris climate conference in December 2015, 195 countries adopted the first-ever universal, legally binding global climate deal.

Research the agreement and create a summary of it for one of the following audiences:

- a 10 year old
- an airline owner
- a farmer.

Include information about:

- problems/challenges
- solutions
- goals/timelines
- how the agreement may affect your audience.

Summative assessment

Statement of inquiry:

Our ability to collect evidence improves with advances in science and technical innovations.

Introduction

In this summative assessment, we will begin by examining evidence of the rising levels of carbon dioxide, a product of human activities. Next, we will design an investigation to model global warming in a closed system. Finally, we will examine scientific data that is evidence of the loss of land ice.

Changes in carbon dioxide levels

Carbon dioxide, CO_2, is released through human activities such as deforestation and burning fossil fuels, as well as natural processes such as respiration and volcanic eruptions. It is a greenhouse gas. The National Oceanic and Atmospheric Administration based in Maryland, USA derived the following information from atmospheric measurements and ice cores.

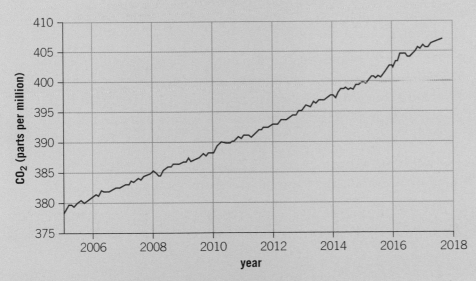

▲ Atmospheric CO_2 levels at Mauna Loa Observatory, Hawaii with seasonal variations removed

▶ Historic CO_2 levels during the last three glacial cycles, as reconstructed from ice cores

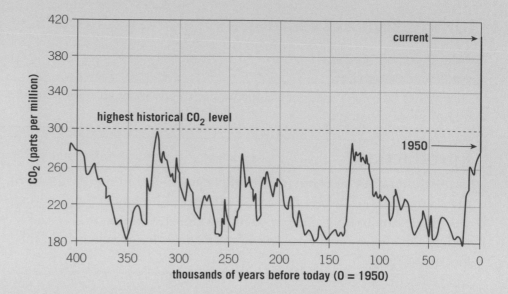

1. Describe the relationship between carbon dioxide concentration and time. [1]

2. Calculate the percentage increase in carbon dioxide concentration in the period 2006–2016. [2]

3. Look at the historic pattern of carbon dioxide atmospheric concentration levels. Describe the most significant aspect of present-day levels of carbon dioxide. [3]

4. How is the global community attempting to limit the increase in carbon dioxide emissions? [3]

5. The alkane, hexane, C_6H_{14}, can exist as five different structural isomers. An isomer is a compound that has the same molecular formula but a different structural formula. For example, butane can exist as a straight chain or as the isomer 2-methylpropane. Both these isomers have the molecular formula C_4H_{10}.

   Apply your knowledge of nomenclature rules for the naming of alkanes and determine the structure and name of the five different isomers of hexane. [10]

 Modelling global warming

Global warming can be defined as the actual and predicted increases in temperature of the Earth's atmosphere and oceans. The predictions are based on many types of evidence collected around the world by the scientific community.

Consider how you could use the following equipment to model global warming and the effect of the increasing amount of greenhouse gases in the atmosphere:

● three digital temperature probes/software or three glass thermometers

● three 1.5 litre soda bottles

- rubber bungs with a hole for the temperature probes/thermometers

- distilled water

- antacid tablets (when these are dissolved in water they produce carbon dioxide)

- high-capacity heat lamp

- 1 m ruler

- stopwatch.

6. Formulate a testable hypothesis. [3]

7. Design an experiment to test your hypothesis. The method should include:

- the independent and dependent variables, and other variables being controlled

- how you plan to record your quantitative and qualitative observations. [10]

 Glaciers and climate change

Land ice sheets at the poles and glacial ice at high altitudes have been the subject of extensive research and monitoring as they provide evidence of accelerating global warming. Land ice and glacial ice do not contribute to sea levels in the way icebergs do. However, when melted, the water joins the planet's oceans, resulting in rising sea levels.

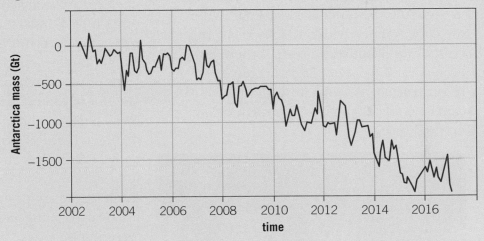

▲ Antartica mass variation since 2002

Source of data: Ice mass measurements by NASA's GRACE satellites (https://climate.nasa.gov/vital-signs/land-ice/)

8. This graph shows the Antarctica mass variation since 2002.

a) Interpret the presented data and describe the trend over the past 16 years. [3]

b) Apart from the overall trend in the data, hypothesize why there are small upward and downward variations in the total land ice mass within a calendar year. [2]

9. a) Estimate the change in mass for the period 2004–2008. [1]

b) Analyze the data and comment whether the rate of loss of mass is increasing or decreasing, in the decade following 2008. Support your argument with mathematical evidence. [4]

This article is from the website of the National Snow and Ice Data Center, situated in Colorado, USA.

Glacial ice can range in age from several hundred to several hundreds of thousands years, making it valuable for climate research. To see a long-term climate record, scientists can drill and extract ice cores from glaciers and ice sheets. Ice cores have been taken from around the world, including Peru, Canada, Greenland, Antarctica, Europe, and Asia. These cores are continuous records providing scientists with year-by-year information about past climate. Scientists analyze various components of cores, particularly trapped air bubbles, which reveal past atmospheric composition, temperature variations, and types of vegetation. Glaciers preserve bits of atmosphere from thousands of years ago in these tiny air bubbles, or, deeper within the core, trapped within the ice itself. This is one way scientists know that there have been several Ice Ages. Past eras can be reconstructed, showing how and why climate changed, and how it might change in the future.

▲ These glaciers in the Himalaya Mountains of Bhutan have been receding over the past few decades, and lakes have formed on the surfaces and near the termini of many of the glaciers

Scientists are also finding that glaciers reveal clues about global warming. How much does our atmosphere naturally warm up between Ice Ages? How does human activity affect climate? Because glaciers are so sensitive to temperature fluctuations accompanying climate change, direct glacier observation may help answer these questions. Since the early twentieth century, with few exceptions, glaciers around the world have been **retreating** at unprecedented rates. Some scientists attribute this massive glacial retreat to the Industrial Revolution, which began around 1760. In fact, several **ice caps**, glaciers and **ice shelves** have disappeared altogether this century. Many more are retreating so rapidly that they may vanish within a matter of decades.

Scientists are discovering that production of electricity using coal and petroleum, and other uses of fossil fuels in transportation and industry, affects our environment in ways we did not understand before. Within the past 200 years or so, human activity has increased the amount of carbon dioxide in the atmosphere by 40 percent, and other gases, such as methane (natural gas) by a factor of 2 to 3 or more. These gases absorb heat being radiated from the surface of the earth, and by absorbing this heat the atmosphere slowly warms up. Heat-trapping gases, sometimes called "greenhouse gases," are the cause of most of the climate warming and glacier retreat in the past 50 years. However, related causes, such as increased dust and soot from grazing, farming, and burning of fossil fuels and forests, are also causing glacier retreat. In fact, it is likely that the earliest parts of the recent glacier retreats in Europe were caused by soot from coal burning in the late 1800s.

The 1991 discovery of the 5,000-year-old "ice man," preserved in a glacier in the European Alps, fascinated the world (see *National Geographic*, June 1 1993, volume 183, number 6, for an article titled "Ice Man" by

David Roberts). Tragically, this also means that this glacier is retreating farther now than it has in 5,000 years, and other glaciers are as well. Scientists, still trying to piece together all of the data they are collecting, want to find out whether human-induced global warming is tipping the delicate balance of the world's glaciers.

10. Explain why glacial ice, and the air bubbles trapped within it, is such an important tool in the study of climate. Include details of the various types of information scientists can ascertain from this scientific endeavour. [3]

11. What modelling can be undertaken to help scientists in their future work? [2]

12. How does this type of information support government and the United Nations' actions? [2]

13. The 20th century has been a time of rapid glacial retreat. What do some scientists attribute this to? [2]

14. Create a detailed diagram to explain the roles played by humans, manufacturing, transportation and public infrastructure, agriculture, fossil fuel energy production and other similar factors in causing global warming and glacial retreat. [6]

3 Consequences

The gastric pits visible in this scanning electron micrograph of the lining of the human stomach contain cells that release digestive enzymes and hydrochloric acid. As a consequence of these acidic conditions, the protein-digesting enzyme pepsin is produced from its inactive precursor pepsinogen. How does the stomach protect itself from its acid contents?

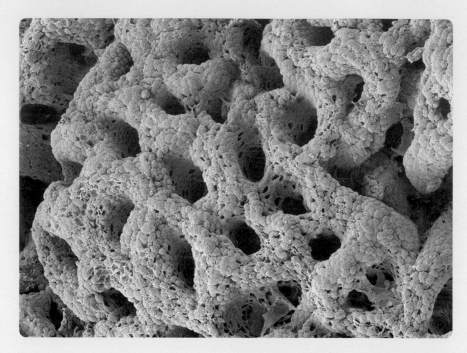

Sinkholes can result from human activity such as mining and construction. Bedrock, the rock layer found beneath the upper soil layer, is often composed of limestone. Calcium carbonate found in limestone reacts with acidic rainwater, breaking down the rock, and creating cracks and pockets beneath the upper layers. What factors contribute to increased acidity in rainwater?

▶ Lactic acid, $C_3H_6O_3$, is produced in muscles from carbohydrates under low-oxygen conditions, for example, when doing vigorous exercise. In industry, lactic acid comes from renewable feedstock such as corn starch or sugar cane. Several lactic acid molecules can be joined together to make one very long molecule, known as polylactic acid (PLA). PLA is a thermoplastic material that is also biodegradable. In what other ways has industry embraced green chemistry?

◀ Blue light, emitted by electronics such as tablets and smartphones, has been shown to affect sleep patterns, as well as the quality of sleep. Research into teenagers' sleep requirements suggests that they need, on average, 8-10 hours of sleep every night. However, many students report sleeping as few as 6 hours per night. This is often a consequence of using smartphones and tablets before going to sleep. How many hours of sleep do you get each night?

Key concept: Change

Related concept: Consequences

Global context: Globalization and sustainability

▼ Salt accumulation in the soil and water can have devastating consequences on the environment: these red gum trees in Southern Australia were killed by the rising salinity in irrigated land. The science behind the human impact on the environment is often well understood, and analyzing the data it provides could help us minimize negative impact. What are the consequences of communities and governments not heeding the warnings?

Introduction

Consequences are defined as the observable or quantifiable effects, results, or outcomes correlated with an earlier event or events. All actions have consequences, both positive and negative; our actions from the past have consequences today and in the same way our actions today will have consequences tomorrow, and some of today's actions (especially collective actions) will have consequences further into the future.

Science has long understood the concept of cause and effect, also known as causality. However, a consequence could be the result of several different factors acting together, meaning it is not always possible to establish causality. For example, the life expectancy of people living in countries such as Japan, Switzerland, Australia, Singapore, Spain and Italy is over 80 years old. This can be attributed to a number of contributing factors including access to clean water, sufficient healthy food and improved access to medicines. While changes in the last half-century have resulted in an overall increase in the quality of life, there have also been less desirable consequences.

Within the scientific community there is little debate about the occurrence of global warming (the increase in the Earth's average surface temperature) or climate change (the long term change in the Earth's climate and patterns of weather), as there is a wealth of scientific evidence proving that these phenomena are occurring. Understanding what causes them and how to mitigate their effects on the future of our planet is a significant concern for current and future generations. The global context of this chapter is globalization and sustainability.

Statement of inquiry:

Change as a consequence of human development can be identified within all environments on our planet.

Since the 19th century, our planet's rapidly developing societies have become increasingly and irreversibly linked to the refining and combustion of fossil fuels. The acidification of our atmosphere is one of the resulting consequences. Non-renewable fossil fuels, such as coal, are burned in power stations on a massive scale to produce electricity for cities and industry. By-products of this combustion reaction include sulfur dioxide, SO_2, and nitrogen dioxide, NO_2. In the atmosphere and under the influence of sunlight, these compounds combine with water to form acid rain and acid snow, which fall back to the surface of our planet.

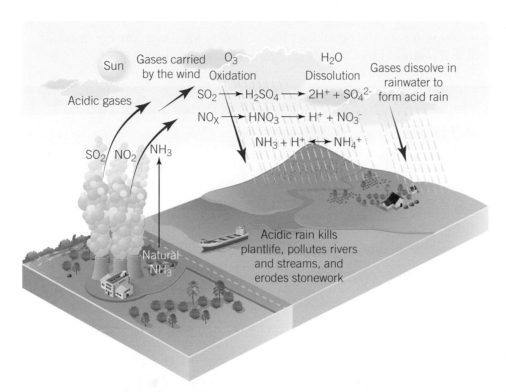

◀ The Earth's atmosphere does not observe the geographical boundaries that exist for different countries. The atmosphere belongs to all countries, its inhabitants and the plant and animal kingdoms. Acid deposition is a global consequence of industrialization and modernization

In the diagram:

Sun

Gases carried by the wind

O_3 Oxidation

H_2O Dissolution

Gases dissolve in rainwater to form acid rain

Acidic gases

$SO_2 \longrightarrow H_2SO_4 \longrightarrow 2H^+ + SO_4^{2-}$

$NO_X \longrightarrow HNO_3 \longrightarrow H^+ + NO_3^-$

$NH_3 + H^+ \longleftrightarrow NH_4^+$

SO_2 / NO_2 NH_3

Natural NH_3

Acidic rain kills plantlife, pollutes rivers and streams, and erodes stonework

ATL Critical thinking skills

Analyzing and evaluating issues and ideas

Often in the media, we read and hear governments, organizations and individuals calling for a reduction in the consumption of non-renewable energy sources, such as fossil fuels. Sometimes people call for a complete ban on the use of carbon-based fuels. Scientists and governments recognise the diversity of products that are made from crude oil, a type of fossil fuel. How many different crude-oil based compounds and their functions can you name?

● Use your critical literacy skills to perform research within a small group, or individually, on the products that are made from crude oil components.

● Construct a list of products and make an informed decision about whether these products are essential or non-essential in society. How could you communicate the level of importance your group placed on each item?

● Discuss within your group the issues that society could face when minimizing or totally eliminating the use of crude-oil products.

Gasoline is the fuel used by the combustion engines that power our cars, buses and trucks. When burned, gasoline and diesel fuel release sulfur dioxide into the atmosphere, as well as carbon monoxide and carbon dioxide. In 2010, there were over 1 billion cars on the road. The Organisation for Economic Co-operation and Development has estimated that the number of vehicles worldwide will reach 2.5 billion by 2050.

Industrialization, modernization and surging population growth have led to ever increasing amounts of pollutants being released into the Earth's atmosphere. Our individual actions, as well as the collective actions of communities that inhabit our planet, always have consequences. Who is taking (or should be taking) responsibility for these consequences of global pollution that are affecting the Earth's atmosphere?

ATL Thinking in context

Rising sea levels

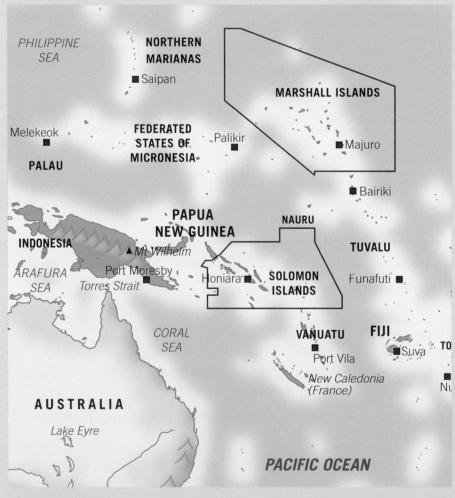

The Solomon Islands and Marshall Islands are experiencing direct consequences of a changing environment due to human development, despite their minimal contributions to global emissions of greenhouse gases. The Green Climate Fund, part of the United Nations Framework Convention on Climate Change, was founded to help developing nations respond to the challenges of our warming planet and to advance the goal of keeping global temperature increase below 2°C. To what extent is it the responsibility of first world nations to aid more vulnerable societies, financially and/or with geographic relocation of residents, in dealing with the unavoidable impacts of climate change?

▲ Both the Solomon Islands and the Marshall Islands, located in the Pacific Ocean, are becoming submerged by rising sea levels

What are acids and bases?

Scientists have devoted decades to discovering, researching and defining two classes of substance—**acids** and **bases**. These groups of compounds and the reactions between them have significant consequences in our everyday lives, as we use and rely on, both directly and indirectly, many substances that are acids or bases.

Your initial understanding of acids and bases has likely been gained from personal experience. For example:

- a lemon tastes sour as a consequence of the presence of citric acid, $C_6H_8O_7$

- adding vinegar, which contains ethanoic acid, $C_2H_4O_2$, gives a little acidity to a salad

- cleaning products work because they contain the weak base ammonia, NH_3

- soil in a garden can be made less acidic by adding ground limestone, which contains calcium carbonate, $CaCO_3$.

The chemical structure of a compound can sometimes be an indicator of the type of compound and its chemical properties.

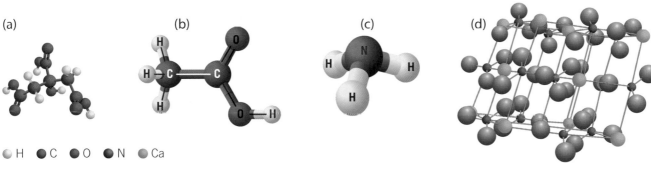

(a)   (b)   (c)   (d)

● H ● C ● O ● N ● Ca

▲ Structures of (a) citric acid, (b) ethanoic acid, (c) ammonia and (d) calcium carbonate

The **pH scale** is a simple method of representing the concentration of hydrogen ions in a solution. A pH value of a solution distinguishes between acidic, neutral and alkaline solutions. We will discuss how to calculate the pH of a solution in the second half of this chapter.

Your stomach contains highly acidic gastric juices, with a pH around 1.0–3.0. In contrast, the liquid inside the duodenum, which is the first section of your small intestine, is slightly alkaline with a pH of 7.4. The term alkali is used to describe a base that is soluble in water.

In chemistry, an acid and a base can be defined in a number of ways. The most widely used definitions are by scientists August Arrhenius, Johannes Brønsted and Thomas Lowry, and G.N. Lewis who each developed their own acid-base theories. In these definitions, the properties of bases also apply to alkalis.

► A wide variety of substances that we use each day are classified as acidic, neutral or alkaline

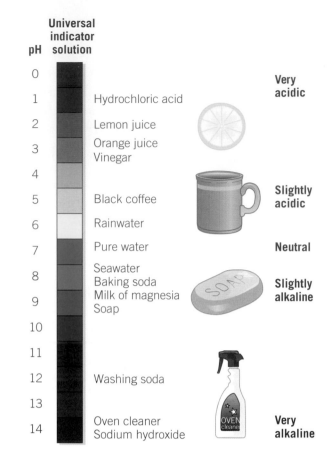

pH	Universal indicator solution		
0			Very acidic
1	Hydrochloric acid		
2	Lemon juice		
3	Orange juice Vinegar		
4			
5	Black coffee		Slightly acidic
6	Rainwater		
7	Pure water		Neutral
8	Seawater Baking soda		Slightly alkaline
9	Milk of magnesia Soap		
10			
11			
12	Washing soda		
13			
14	Oven cleaner Sodium hydroxide		Very alkaline

Arrhenius was awarded the Nobel Prize in 1903 for the research he performed on acids and bases. His definition of an **acid** as a substance that ionizes in water to produce hydrogen ions, H^+, and an **alkali** as a substance that ionizes in water to produce hydroxide ions, OH^-, are the definitions we will use in this book.

Acids and bases have characteristic properties that enable them to be identified in the laboratory, for example:

Property	Acids	Bases
Taste	Sour	Bitter
pH	pH < 7.0	pH > 7.0
Effect on indicators	Turn blue litmus red	Turn red litmus blue
	Phenolphthalein - colorless	Phenolphthalein - pink
	Methyl orange - red	Methyl orange- yellow

Extreme aquatic environments

Soda lakes are naturally occurring alkaline lakes characterized by a pH of 9–12 which results from high concentrations of carbonate salts. These lakes often contain high concentrations of dissolved salts such as sodium chloride, which also makes them saline. This combination of salinity and alkalinity makes soda lakes some of the most extreme aquatic environments on our planet. These extreme environments harbor a high diversity of microorganisms, including unique species that are adapted to alkaline conditions, called alkaliphiles.

▲ Mono Lake is one of the oldest lakes in North America, with an estimated age of over 1 million years. Located 1,900 m above sea level, it has a very high salinity level compared to seawater

Mono Lake has undergone changes as a result of human development. These changes have had numerous consequences. Referring to the website on Mono Lake (http://www.monolake.org/about/stats), answer the following questions:

1. Identify the main dissolved salts contributing to the alkalinity of the lake.

2. How does the pH of the lake compare to the pH of typical ocean water?

3. Describe how the salinity level of the lake has changed over the past 100 years. To what extent might the lake have changed naturally over time?

4. State what types of organisms can be found in the lake.

5. Comment on how the diversion of water for an increasing population of Los Angeles has affected the salinity of the lake.

6. What impact has human development and intervention had on the ecosystem of the lake?

What is the difference between strong and weak acids or bases?

INORGANIC

What comes to mind when you hear the word acid? Most people think about a dangerous corrosive liquid used in laboratories and industry. Not all acids would match that perception – as we have seen, acids occur in foods such as vinegar and fruit juice. The scientific community understands the importance of explaining and educating society on the accurate use of scientific terms such as "strong acid". This understanding helps empower people to examine claims in the media and determine if those claims are valid.

In general, the difference between a strong acid or base and a weak one is the degree to which it **dissociates**, or breaks up into ions in water. Some examples of acids and bases will help illustrate this concept.

Strong acids are corrosive by nature. Acids will chemically destroy living tissue and will react with and break down a wide variety of substances such as metals and organic compounds. Hydrochloric acid is defined as a strong acid—it fully dissociates in water. When dissolved in water, the chloride ion has no affinity for the hydrogen ion. In an aqueous solution, the chloride ion, Cl^-, and the hydrogen ion, H^+, exist as separate ions.

$$HCl(aq) \rightarrow H^+(aq) + Cl^-(aq)$$

The dissociation reaction for a **weak acid** reaches equilibrium after only a small proportion of the acid has dissociated in solution. Ethanoic acid, CH_3COOH, is commonly known as acetic acid; it is a weak acid found in vinegar. An equilibrium exists when ethanoic acid partially dissociates in water. The ethanoate ion, CH_3COO^-, has a strong affinity for the hydrogen ion, H^+.

$$CH_3COOH(aq) \rightleftharpoons H^+(aq) + CH_3COO^-(aq)$$

The equilibrium position here favours the reactant side. Consider what this means in terms of concentration of the reactants and products in solution.

Strong bases, like strong acids, completely dissociate in water while a weak base partially dissociates in water to form alkaline solutions. Metal hydroxides such as sodium hydroxide, $NaOH$, are examples of strong bases. Sodium hydroxide completely dissociates in water, and in an aqueous solution, the hydroxide ion, OH^-, has no affinity for the sodium ion, Na^+.

$$NaOH(aq) \rightarrow Na^+(aq) + OH^-(aq)$$

▶ Behavior of strong and weak acids in aqueous solution

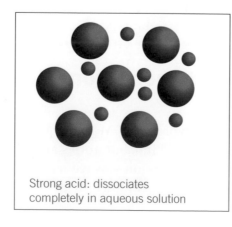

Strong acid: dissociates completely in aqueous solution

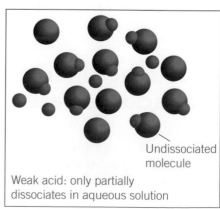

Undissociated molecule

Weak acid: only partially dissociates in aqueous solution

Strong bases are **caustic** by nature. An alkaline solution will corrode, burn or destroy living tissue. While chemists describe strong acids as corrosive chemicals and strong bases as caustic substances, both must be handled very carefully.

Ammonia is an example of a **weak base**. Ammonia is made industrially in very large quantities using the Haber process and is one of the most versatile chemicals produced globally. Ammonia reacts with water, and the reaction reaches equilibrium.

$$NH_3(aq) + H_2O(l) \rightleftharpoons NH_4^+(aq) + OH^-(aq)$$

What is the difference between strength and concentration?

INORGANIC

The *strength* of an acid or a base is different from the *concentration* of an acidic or basic solution. Solutions that are **concentrated** contain a high number of particles of the compound for a given volume of solvent.

Imagine a glass of chocolate milk. A drink containing a large amount of cocoa powder is described as **concentrated** rather than strong. If there is very little cocoa powder in the drink, it is described as **dilute**.

Acids and bases that are used in the laboratory may be strong or weak. They can each be prepared as dilute or concentrated solutions:

	Dilute solution	Concentrated solution
Weak acid or base	Ethanoic acid 0.5 mol dm^{-3}	Ammonia 6.0 mol dm^{-3}
Strong acid or base	Sodium hydroxide 0.5 mol dm^{-3}	Sulfuric acid 6.0 mol dm^{-3}

▲ A solution of high concentration is made up of a large amount of solute (solid) for a given volume of solvent

 Experiment

What is the reaction between a metal carbonate and an acid?

Your task is to collect the gas that is produced in the reaction between calcium carbonate and hydrochloric acid. You will then perform a chemical test to identify the composition of this gas. Data collected should be organized and presented appropriately, and then analyzed to infer a valid conclusion.

⚠ **Safety**

● Acids are corrosive and should be handled safely. Avoid contact with the skin and inhalation of the vapors.

● Safety glasses must be worn at all times.

Materials

- Calcium carbonate (marble chips)
- 1 mol dm⁻³ hydrochloric acid
- Calcium hydroxide solution (limewater)
- A boiling tube
- A test tube
- 25 cm³ measuring cylinder

There are a number of different ways to set up the apparatus needed for this experiment; the diagram shows one of these alternatives.

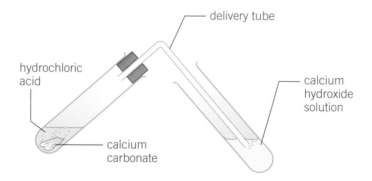

hydrochloric acid

delivery tube

calcium hydroxide solution

calcium carbonate

▲ Apparatus used for the reaction between calcium carbonate and hydrochloric acid

Method

1. Half fill the test tube with calcium hydroxide solution and place into a rack for later.

2. Place 1.0–1.5 g of calcium carbonate in the boiling tube, then place into the rack.

3. Using the measuring cylinder, measure 15 cm³ of 1 mol dm⁻³ hydrochloric acid and transfer it to the boiling tube. Seal the tube with a rubber stopper that has the delivery tube attached to it.

4. Remove the test tube containing the calcium hydroxide solution from the rack and place it so that the free end of the delivery tube is below the surface of the solution.

5. Record any qualitative observations.

Questions

1. Write the balanced chemical equation for the reaction between calcium carbonate and hydrochloric acid. Can you predict the gas produced from the composition of the reactants?

2. Describe what you saw happening to the calcium hydroxide solution.

3. Write a balanced chemical equation for the reaction between calcium hydroxide solution and carbon dioxide gas.

Battery disposal

Each year, billions of batteries of all types are bought, used and discarded. Some of these batteries, such as alkaline batteries, use potassium hydroxide, KOH, as an electrolyte; others, such as zinc-carbon batteries, use an electrolyte of zinc chloride, $ZnCl_2$. Other types of batteries contain lead and sulfuric acid or nickel and cadmium, some of which are toxic to humans and the environment.

Improper disposal of used batteries can have negative environmental consequences. Commonly, if used batteries are disposed of in recycling containers, local government authorities will treat this as hazardous waste. Disposal methods, such as sealing the batteries in cement-encased steel drums, attempt to prevent the batteries being dumped in landfill and releasing the chemicals into the environment.

◀ Battery recycling collection bins are a common site in many countries. Governments encourage environmentally-safe practices in this way

1. How can some consequences (like improper battery disposal) have far-reaching effects on the environment?

2. Explain how the environment would be affected by the disposal of used batteries into landfill.

3. What would the consequences for our planet be, if plastic waste could be eliminated or significantly reduced?

How does the pH scale work?

INORGANIC

The philosophical principle of Occam's razor is used as a guide to the development of scientific theories. When two different explanations are possible for a theory, the explanation that avoids or has very few assumptions is always preferred. For example, it is better to choose an explanation presenting complex concepts in a simple, straightforward way, in order to make them accessible to a wider population.

The pH scale is a simple mathematical scale that is easy to understand and use. It enables us to quickly tell how acidic or alkaline a substance is based on its position on the scale. The pH scale is logarithmic, meaning it can represent a very wide range of concentrations of hydrogen ions that would otherwise be difficult to visualize.

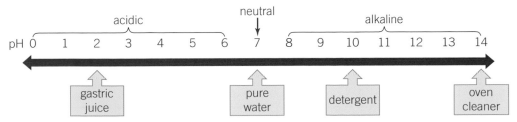

On this scale, pH values below 7 describe acidic solutions. The lower the number, the more acidic and corrosive the acid becomes. A pH value of 7 represents a neutral substance—it is neither acidic, nor alkaline. Numbers above 7 represent alkaline solutions: the higher the number, the more alkaline and caustic the solution.

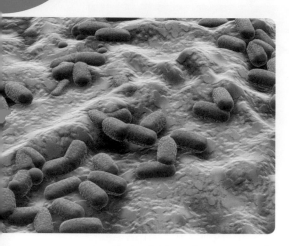

▲ *Escherichia coli*, one type of bacterium found in gastric fluid

Antibiotics, pH levels and gut health

pH is more than just a scale used to characterize solutions; it also offers us valuable information about our health. There is evidence that the diversity of microbes living in our gut has consequences for the modulation of our body's immune system. pH is a determining factor in how effectively gut microbes function, and gastric fluid with elevated pH levels (above 6.0) has been shown to contain an altered composition and reduced diversity of microbiota.

Broad-spectrum antibiotic medications have transformed modern medicine and saved millions of lives, but have also been shown to increase gastric fluid pH level, influencing gut microbe diversity and composition.

1. Thinking about the environment of our gut microbiome, to what extent should we consider the potential effects of over-prescribing antibiotics in relation to our health?

2. To what extent are these health consequences heightened in immunocompromised individuals? Would individuals with weaker immune systems be affected more or less? Why?

3. How might the over-prescription of antibiotics have effects that extend beyond our gut environment?

4. Evaluate and justify to what extent the benefits of antibiotic medication overshadow the consequences of their over-use.

INORGANIC

How is the pH of a solution calculated?

The pH scale is a logarithmic scale to base 10. For every change of 1 pH unit, there is a 10-fold change in the dissolved hydrogen ion concentration. This is best understood by considering the comparison of the pH of substances relative to water.

Substance	pH	Concentration of hydrogen ions (mol dm^{-3})	Change in hydrogen ion concentration compared to water
Water	7.0	1.0×10^{-7}	0
Milk	6.0	1.0×10^{-6}	$\times 10$
Black coffee	5.0	1.0×10^{-5}	$\times 100$

The pH of a solution is calculated using the expression:

$$pH = -\log_{10}[H^+]$$

Data-based question: Comparing acidity

pH	Example	pH	Example
1	Gastric juice	9	Baking soda
5	Black coffee	13	Bleach
7	Water	14	Liquid drain cleaner

1. How can you determine which is the more acidic solution, black coffee or gastric juice?

2. Can you deduce how many times greater the concentration of hydrogen ions is in the more acidic of the two solutions?

3. Black coffee and baking soda are both 2 pH units away from water. Black coffee is acidic and baking soda is alkaline. How do their hydrogen ion concentrations compare?

4. Gastric juice is a corrosive liquid and bleach is a caustic liquid. Explain what these terms mean and how the hydrogen ion concentrations compare.

What is the role of an indicator in acid-base chemistry?

INORGANIC

An indicator is a weak acid or a weak base that displays a characteristic color, depending on the pH of the solution. Universal indicator (UI) solution is versatile in that it enables us to determine the specific pH of a solution within the full range of the pH scale.

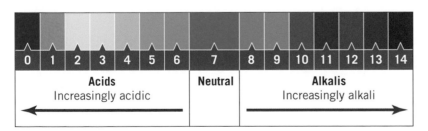

◀ Universal indicator is a mixture of indicators, allowing for a specific pH to be determined from a color chart. UI is useful in general chemistry, but would never be used as the indicator for an acid-base titration

Indicators generally behave as weak acids, establishing an equilibrium in solution. The equilibrium position is pH dependent.

$$HIn(aq) \rightleftharpoons H^+(aq) + In^-(aq)$$

$$\text{color A} \qquad\qquad \text{color B}$$

Le Chatelier's principle can be applied to help understand how the color of an indicator is pH dependent. In acidic solutions, the hydrogen ion is in excess and the equilibrium position favours the left-hand side of the equation. The HIn form of the indicator becomes dominant in the equilibrium and color A is observed. In basic solutions, the hydrogen ions are depleted when they react with the hydroxide ions.

$$H^+(aq) + OH^-(aq) \rightarrow H_2O(l)$$

The equilibrium position shifts to the right and the In⁻ form of the indicator becomes dominant, and color B is observed.

The choice of indicator in an acid-base titration is dependent on the strength of the acid and base. An example of some common indicators and the acid-base titration they are used for is shown in the table below. You will study indicator theory more deeply in the IB DP Chemistry course.

Indicator	Color in acid solution	Color in basic solution	Titration
Phenol red	Yellow	Red	Strong acid-strong base
Methyl orange	Red	Yellow	Strong acid-weak base
Phenolphthalein	Colorless	Pink	Weak acid-strong base

INORGANIC # What is a neutralization reaction?

The **neutralization** of an acid by a base is an example of an **exothermic reaction**. Heat is generated in the reaction and is transferred to the surroundings. In a neutralization reaction, a **salt** is produced, made up of a **cation** (positive ion) from the base and an **anion** (negative ion) from the acid. Water is a by-product in this reaction.

The word equation for all acid-base neutralization reactions is:

$$acid + base \rightarrow a\ salt + water$$

The composition of the acid and base determine which salt is formed:

- hydrochloric acid produces chloride salts
- sulfuric acid produces sulfate salts
- nitric acid produces nitrate salts.

For example, in the reaction between hydrochloric acid and sodium hydroxide, a chloride salt (sodium chloride) will be formed:

$$HCl(aq) + NaOH(aq) \rightarrow NaCl(aq) + H_2O(l)$$

Formation of salts

Now that you know how salts are formed from the reaction between an acid and a base, complete the following word equations:

1. Hydrochloric acid + lithium hydroxide →

2. Nitric acid + potassium hydroxide →

3. Sulfuric acid + calcium hydroxide →

4. _____ + sodium hydroxide → sodium sulfate + _____

Which metals react with acids?

The reactivity series is a list of metals arranged from most reactive to least reactive. Reactive metals like sodium and magnesium are found towards the top of the series. They have a tendency to lose electrons and form cations. Metals at the bottom of the series such as gold and platinum are often referred to as precious metals. They are found in their "native" state, unreacted with any other element. They do not lose electrons readily.

Metals found above hydrogen in the reactivity series will react with an acid to form a salt and hydrogen gas.

The reactivity series

Potassium	**Most reactive**
Sodium	
Calcium	
Magnesium	
Aluminium	
(Carbon)	
Zinc	
Iron	
Tin	
Lead	
(Hydrogen)	
Copper	
Silver	
Gold	
Platinum	**Least reactive**

◀ The reactivity series is a description of the reactivity of metals and select non-metals relative to one another. It enables chemists to predict the outcome of displacement reactions and redox reactions

▲ The reaction between magnesium and hydrochloric acid is vigorous. How would your observations change if magnesium was replaced with a metal lower in the reactivity series?

Magnesium is above hydrogen in the reactivity series, so it reacts with hydrochloric acid to form the salt, magnesium chloride:

Magnesium + hydrochloric acid → magnesium chloride + hydrogen

$$Mg(s) + 2HCl(aq) \rightarrow MgCl_2(aq) + H_2(g)$$

Experiment

How do you test for the presence of hydrogen gas?

The hydrogen test is also known as the pop test. In this simple test, hydrogen gas combines with oxygen gas present in the air and is ignited. The combustion reaction creates a small scale explosion, heard as a "pop", and water vapor is the product.

Safety

- Wear gloves when performing this experiment.
- Hydrochloric acid is corrosive. Safety glasses must be worn at all times.

Materials

- 1 mol dm^{-3} hydrochloric acid
- Magnesium ribbon
- Test tube
- Rubber stopper
- Wooden splint

Method

1. Transfer into a test tube using a dropping pipette 5 cm^3 of 1.0 mol dm^{-3} hydrochloric acid.

2. Place a 2 cm long piece of magnesium ribbon into the test tube and seal the test tube with a rubber stopper. You may need to hold the stopper to prevent it popping off the test tube as the gas pressure builds.

3. Light a wooden splint and when you decide that there is sufficient gas pressure, remove the stopper and place the lit splint at the mouth of the test tube.

Questions

1. Energy is conserved in all chemical reactions. Name one form of energy that you experience in this reaction.

2. Write a balanced chemical equation for this reaction.

What are the products of the combustion of fossil fuels?

Fossil fuels include crude oil, coal and natural gas. Crude oil or petroleum is a non-renewable resource and its use as a fuel has become deeply embedded within our global society. Crude oil is a natural mixture of **hydrocarbons**, which are chemicals that contain carbon and hydrogen only, and organic compounds containing nitrogen, sulfur and oxygen, and a variety of metals.

The combustion of fossil fuels releases energy, which is the reason for carrying out the reaction on a massive scale. Other products of the reaction are carbon dioxide and water.

Coal, petroleum and natural gas are the main fuels used to generate energy, for example, in power stations or the internal combustion engines of cars. As the global demand for energy increases, so does the consumption of fossil fuels. A consequence of this is the release of large quantities of carbon dioxide into the atmosphere, which over many years has led to **global warming**, arguably the most significant challenge facing our planet in human history.

Carbon dioxide emissions

In an attempt to reduce the consequences of burning fuels, alternative energy sources such as ethanol and hydrogen are increasingly being used. In theory, it is possible to rely only on clean, renewable energy sources to satisfy the world's energy requirements, but in practice only about 14% of the total generated energy relies on these.

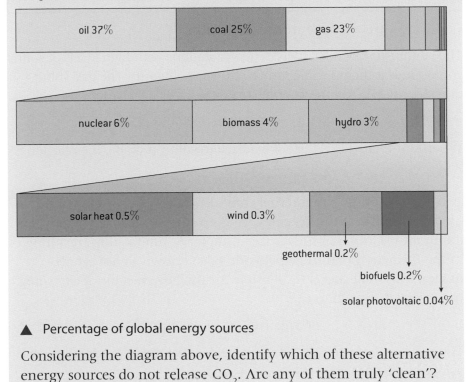

oil 37% | coal 25% | gas 23%

nuclear 6% | biomass 4% | hydro 3%

solar heat 0.5% | wind 0.3%

geothermal 0.2%

biofuels 0.2%

solar photovoltaic 0.04%

▲ Percentage of global energy sources

Considering the diagram above, identify which of these alternative energy sources do not release CO_2. Are any of them truly 'clean'?

▲ Zeppelins were lighter-than-air aircrafts developed at the turn of the 20th century. Once considered to be the future of air travel, they fell out of fashion following the Hindenburg disaster in 1937, which saw a hydrogen-filled Zeppelin explode as it was trying to land

What effect does acid deposition have on the environment?

ENVIRONMENTAL

Acid deposition results from the release of acid-forming pollutants into the atmosphere and subsequent deposition of these substances onto the Earth's surface. Airborne pollutants originate in many countries across the planet, but the action of these pollutants has no geographical boundaries and air pollution is an international problem. Compounds that lead to acid deposition, such as sulfuric acid and nitric acid, are regarded as secondary pollutants, while the carbon dioxide produced in the combustion of fossil fuels is a primary pollutant. Secondary pollutants are formed when primary pollutants react with other chemical compounds found in the atmosphere.

Sulfur dioxide, SO_2, and nitrogen oxides, NO_x, are both primary pollutants. When they are emitted into the atmosphere, they react with water, oxygen and other chemicals to form sulfuric and nitric acids. These strong acids then fall to the ground as acid rain or snow.

While the burning of fossil fuels is a major source of pollutants linked to the formation of acid rain, a small proportion of these compounds come from natural sources:

Natural occurrence	Gas produced	Acid produced
Gases released from active volcanoes	SO_2	H_2SO_3 / H_2SO_4
Decomposing vegetation	CO_2	H_2CO_3
Electrical discharge (lightning)	NO	HNO_3

Can we control these sources of pollutants?

The combustion of fossil fuels in power stations with high levels of sulfur compounds is the principle source of these pollutants. The gas, sulfur dioxide, combines with water to create the weak acid, sulfurous acid:

$$SO_2(g) + H_2O(l) \rightleftharpoons H_2SO_3(aq)$$

When sulfur dioxide is further oxidized to sulfur trioxide, the reaction is:

$$SO_3(g) + H_2O(l) \rightleftharpoons H_2SO_4(aq)$$

The burning of fuels in internal combustion engines produces nitric oxide, (NO). The NO molecules are then further oxidized by oxygen donor atoms found in the air (see Chapter 8, Interaction), resulting in the formation of nitrogen dioxide (also known as nitrogen(IV) oxide). This molecule is responsible for the brown color of city smog. In general the reaction is:

$$2NO(g) + O_2(g) \rightarrow 2NO_2(g)$$

▲ Mexico City is an example of a mega city that has high levels of pollution. The brown smog is evidence of nitrogen(IV) oxide in the atmosphere. The consequences of the presence of this gas in the atmosphere range from irritation of the airways in the human respiratory system, to acid rain

The reaction between water and nitrogen(IV) oxide produces nitric acid and nitrous acid:

$$2NO_2(g) + H_2O(l) \rightarrow HNO_3(aq) + HNO_2(aq)$$

Acid deposition has global implications, not just for the countries producing the pollutants. National monuments and historic buildings throughout the world that are made from limestone or marble are subject to erosion.

Acid deposition also results in lakes and streams becoming more acidic. Acid rain can remove essential nutrients from the soil, depleting the soil of the requirements for growth of trees and plants leading to the death of forests on a large scale. Acidic rain reacts with aluminium compounds in the soil and dissolves aluminium ions, carrying them into the aquatic environment. Dissolved aluminium is poisonous to some species of fish. Other marine life can only tolerate small changes in the pH of their ecosystem. When the changes in pH become more significant, spawning and the hatching of eggs can be affected and species can die, often in large numbers.

◄ Calcium carbonate is one of the main constituents of limestone. It undergoes a neutralization reaction with an acid, releasing carbon dioxide and resulting in the weakening of buildings and monuments

Summative assessment

Statement of inquiry:

Change as a consequence of human development can be identified within all environments on our planet.

Introduction

In this summative assessment, you will first examine the phenomenon of acid rain. Then you will design investigations that simulate acid rain and the neutralization of acidic soils. Finally, you will perform a case study on the effects of acid rain on terrestrial or aquatic systems.

The article below contains an explanation of acid deposition found on the United States Environmental Protection Agency (EPA) website at https://www.epa.gov/acidrain/what-acid-rain. Read the passage carefully then complete the tasks that follow.

What is acid rain?

Acid rain, or acid deposition, is a broad term that includes any form of precipitation with acidic components, such as sulfuric or nitric acid that fall to the ground from the atmosphere in wet or dry forms. This can include rain, snow, fog, hail or even dust that is acidic.

Wet and dry deposition

Wet deposition is what we most commonly think of as acid rain. The sulfuric and nitric acids formed in the atmosphere fall to the ground mixed with rain, snow, fog, or hail.

Acidic particles and gases can also deposit from the atmosphere in the absence of moisture as *dry deposition*. The acidic particles and gases may deposit to surfaces (water bodies, vegetation, buildings) quickly or may react during atmospheric transport to form larger particles that can be harmful to human health. When the accumulated acids are washed off a surface by the next rain, this acidic water flows over and through the ground, and can harm plants and wildlife, such as insects and fish.

The amount of acidity in the atmosphere that deposits to Earth through dry deposition depends on the amount of rainfall an area receives. For example, in desert areas the ratio of dry to wet deposition is higher than an area that receives several inches of rain each year.

▲ Dry deposition sampling is occurring on a global scale

 ## Measuring acid rain

1. Normal rain has a pH of about 5.6; it is slightly acidic because carbon dioxide (CO_2) dissolves into it forming carbonic acid, a weak acid. Acid rain usually has a pH between 4.2 and 4.4. Acid deposition occurs when sulfur dioxide, SO_2, and nitrogen oxides, NO_x, are emitted into the atmosphere and react with water, oxygen and other chemicals present in the atmosphere.

   a) Write balanced chemical equations for the reactions of sulfur dioxide and nitrogen(IV) oxide with water. [4]

   b) Name the acids produced. [2]

2. Outline the major differences between wet and dry acid deposition. [4]

3. Describe the differences between strong and weak acids. [2]

4. Deduce the chemical reaction for the formation of carbonic acid. Classify this as either a strong or weak acid. [3]

5. Acid rain has a pH of 4.2 to 4.4. This is approximately 1 pH unit lower than normal rain water. Given your knowledge of the pH scale, explain how the concentration of hydrogen ions differs between these two solutions. [2]

Investigating acid rain

Sulfur dioxide, SO_2, is emitted into the atmosphere both naturally (volcanoes) and as a consequence of human activities (combustion of fossil fuels). The mixing of this toxic gas with water is the cause of acid deposition.

6. Design a way of simulating acid rain formation by sulfur pollutants and investigate the effects of this pollutant on the pH of water. You could use the following points for guidance. [10]

> **Equipment**
>
> - sulfur powder
> - sodium hydrogencarbonate
> - universal indicator solution
> - a deflagration spoon
> - bunsen burner
> - gas jar
> - distilled water.
>
> Your design should include:
>
> - how you will generate the sulfur dioxide gas
> - the method used to enable the sulfur dioxide to dissolve in the distilled water
> - how you will test to see if there has been a change in the pH of the solution
> - details of the variables being kept constant during the experiment
> - details of the safety issues and how you will minimize any risks.

7. Propose a research question for your investigation. [2]

8. Formulate a testable hypothesis and explain it using scientific reasoning. [4]

Soil acidity plays an important role in optimizing the growth of plants. The soil acidity can change as a result of acid rain. Reduced pH levels can deplete the soil of essential nutrients and minerals.

9. Having formulated an experiment to replicate the formation of acid rain in the first part of the question, design an experiment to demonstrate how increased soil acidity resulting from acid rain can be diminished or reversed. Your design should include:

- how you will neutralize the effect of the acid rain

- how you will demonstrate that the pH has been changed. [10]

10. **a)** What is the best way to present qualitative observations and quantitative of the type that you would have collected had you performed the investigation you designed in question 6?

b) Work within a small group to evaluate the two experimental methods devised by you and your peers. Identify modifications that could improve the two experiments. [5]

c) Describe relevant improvements and propose any further experiments that could be performed to investigate the effects of acid rain. [5]

Effects of acid rain on terrestrial or aquatic environments

Consequences of human development have posed significant threats to terrestrial and aquatic ecosystems across the globe. Your goal is to write a case study related to the consequences of acid rain on either a forest or a body of water of your choice. The case study should respond to the following questions:

- How can some consequences have far-reaching effects?

- How might changing environments have positive consequences?

Your research should include:

- symptoms of the affected environment

- data that was collected within the environment to determine consequences

- how the effects of acid rain were measured/quantified

- local and global consequences of your chosen environment affected by acid rain

- (for aquatic environments only) sources and consequences related to acid rain, of methylmercury accumulation

- steps that are being taken and/or have been taken in order to reduce human impact on your chosen environment.

Some possible forests to choose from include the Black Forest (Germany) and the Green Mountain Forest (USA); alternatively, if you would like to investigate the consequences of human development on aquatic environments, you could look at the 16 national objectives relating to environmental quality in Sweden.

▲ The Black Forest in Germany has suffered significant damage as a result of acid rain

4 Energy

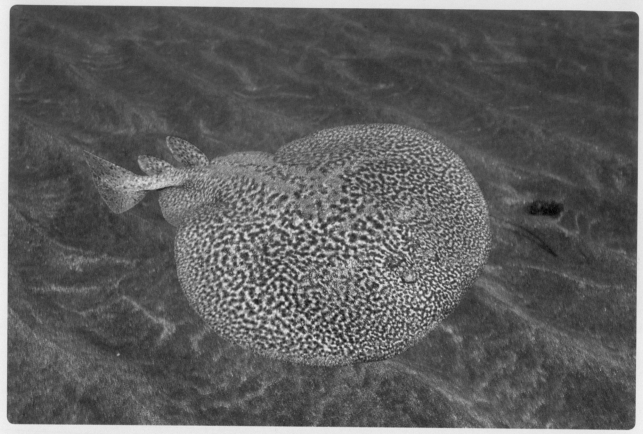

▲ The marbled torpedo ray (*Torpedo marmorata*) is a species of electric ray found in the Eastern Atlantic Ocean. It can produce an electric shock which it uses to subdue its prey and defend itself. The ability of some rays to generate electricity has been known for centuries. The ancient Greeks used the electricity generated to numb the pain of surgery and childbirth

◀ The Oriental hornet (*Vespa orientalis*) lives in colonies. The yellow stripe on its body acts as a solar collector, converting solar energy into electrical energy. The purpose of this is still unknown. The hornets dig their underground nests by picking up soil in their jaws, flying a short distance away, dropping the soil, then returning to the nest to continue work. Solar energy may be contributing to the energy required to do this work. What technology do we use to capture solar energy?

▲ Lightning is a natural phenomenon. Heat energy is transferred from lightning to the surrounding air, which is heated rapidly to over 18,000°C. The rapid expansion of the surrounding atmosphere results in a pressure wave that is heard as a thunder clap. How does lightning result in the formation of nitrate ions, a very useful natural fertilizer?

▲ Icebreakers are used to keep shipping channels clear for other ships and in scientific expeditions in the Arctic. The vessels require powerful engines to drive their way through ice. This nuclear-powered icebreaker is fueled by the heat energy generated when uranium atoms are split. Nuclear-powered icebreakers can only travel in cold waters as they need cold water for their engine cooling system. How else do we use the energy generated by nuclear reactions? What are the risks associated with nuclear energy?

Introduction

Energy is integral to all aspects of our lives. Work is performed as energy is transformed from one type to another. Thermodynamics investigates the transfer of energy from one place to another, and from one form to another. In chemistry we are interested in the interrelation of heat and work in chemical reactions or with physical changes of state.

An example of how we use energy to perform work is our use of fossil fuels to generate electricity, and in turn cook our food. In this example it is clear how energy is being used to perform work: chemical energy stored within fossil fuels is converted into thermal energy and in turn, electrical energy is generated. Other processes are not as obvious. For example, we cannot directly observe the work being performed inside the leaves of plants by light energy from the Sun, but we can observe the products of this work in the growth of plants. For this reason, the key concept of this chapter is change.

Key concept: Change

Related concept: Energy

Global context: Scientific and technical innovation

▼ The striking surface of a matchbox is made of sand, powdered glass and red phosphorus. A match head contains powdered glass, potassium chlorate, sulfur and gelatin. When the match head is struck against the matchbox, the sand and powdered glass cause friction and heat, and this converts some of the red phosphorus to white phosphorus. This chemical is so volatile that it ignites in the air. The heat also breaks down the potassium chlorate in the match head which releases oxygen. The oxygen combines with sulfur and keeps the flame burning. The gelatin acts as glue to keep everything together in the match head, and also provides extra fuel. What are the energy changes that occur in these reactions?

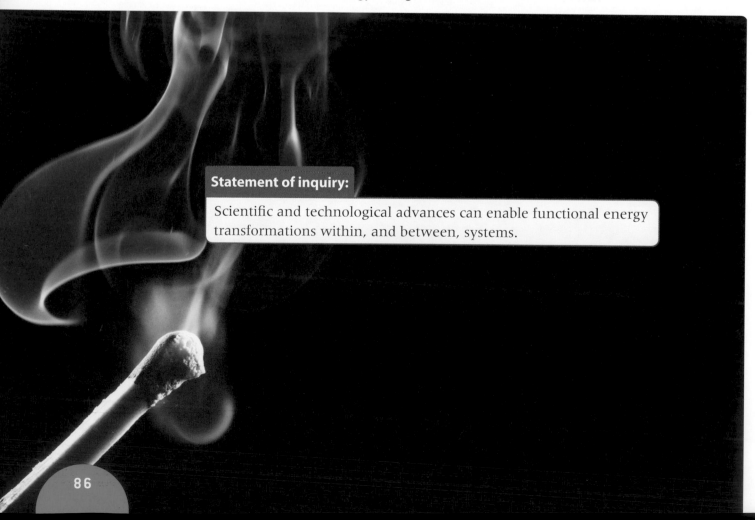

Statement of inquiry:

Scientific and technological advances can enable functional energy transformations within, and between, systems.

Human interactions within communities and our relationship with the natural world are influenced by scientific and technological advances as we seek to fulfill our energy needs. The level of efficiency has increased and cost of solar panels has dropped significantly as a result of technological advancement. For this reason the global context of this chapter is scientific and technical innovation.

▲ This bus is powered by hydrogen fuel cells. This technology has significantly advanced since first being used in space exploration capsules in the 1960s. Hydrogen fuel cells efficiently convert chemical energy into electrical energy. A flow of electrons produced by the cell can be used to do work. Technology is making this energy source more efficient and cost effective, resulting in hydrogen-powered public transport and private vehicles becoming more commonplace. How is clean energy helping the global environment?

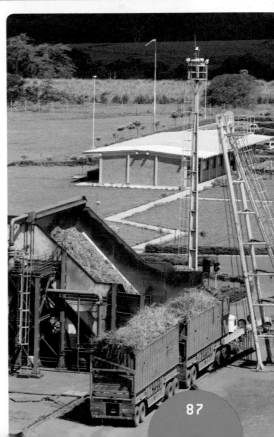

▶ This photograph shows sugarcane being unloaded at a biofuel and sugarcane factory. Farmers in Brazil have grown sugarcane for sugar manufacturing for centuries and it is a major domestic and export crop. Light energy from the Sun is converted into chemical energy and stored inside the sugarcane plant. Bagasse, the residue from processing sugarcane, is the energy source for bioelectricity made at the sugarcane mills. Today, Brazil is one of the largest producers of biofuels, specifically ethanol made from sugarcane. Ethanol combustion converts chemical energy into thermal energy and finally mechanical energy. Do you know any other countries that are using ethanol as an alternative fuel?

What is the role of energy in chemical reactions?

All chemical reactions involve an energy change. Energy may be released into the surroundings or it may be absorbed from the surroundings. Most commonly the energy is in the form of heat, but it may also be as sound or light. Reactions that release energy are referred to as exothermic, while reactions that absorb energy are endothermic.

1. These hot and cold compresses contain a chemical and an inner bag containing water. By squeezing the outer bag, the inner bag is ruptured and the water mixes with the chemical. The chemical in the hot compress is magnesium sulfate; the cold compress contains ammonium nitrate.

   Which reaction is endothermic and which is exothermic?

2. Describe the energy changes that occur after a sparkler is lit.

What happens in an exothermic reaction?

When a metal is added to an acid at room temperature, the temperature of the mixture increases. This is an example of an exothermic (heat-releasing) reaction.

  Experiment

What happens when magnesium reacts with hydrochloric acid?

⚠ Safety

- Wear safety glasses.
- Hydrochloric acid is corrosive. Avoid contact with the skin.

Materials

- 1 mol dm^{-3} of hydrochloric acid
- Strips of magnesium ribbon
- Boiling tube and rack
- Pipette
- Temperature probe or thermometer

Method

1. Place a boiling tube in a rack. Using a pipette, add 10 cm^3 of 1 mol dm^{-3} hydrochloric acid to it.
2. Measure the initial temperature of the acid.
3. Place two 1 cm long strips of magnesium metal into the boiling tube.
4. Record what happens.
5. When the reaction stops record the final temperature of the solution.

Questions

1. How did you know that a reaction had taken place? Give at least two reasons.
2. How did you know that the reaction had finished?
3. Explain why the reaction is endothermic or exothermic.
4. Write the balanced chemical equation for this reaction.

Thermite reaction

Some exothermic reactions produce large amounts of energy that are utilized in the reaction. The thermite reaction between iron(III) oxide and aluminium powder is a redox (reduction–oxidation) reaction that releases an enormous amount of heat energy.

Search online for demonstrations of the thermite reaction. Watch at least two of these demonstrations.

Questions

1. Describe the reaction. What energy changes did you observe?

2. Write the balanced chemical equation for the reaction between iron(III) oxide and aluminium.

The thermite reaction has been used since the 1930s to repair railway and tram tracks. The powdered aluminium and iron(III) oxide are put in a container and lit. The heat from the reaction melts the tracks and the molten metal fills any gaps.

3. What are the advantages of this reaction for railway construction?

4. A thermite reaction can occur if rusty steel is covered by aluminium paint and hit by a hammer. Explain why this might happen.

There are many other examples of exothermic reactions:

- combustion of fuels for powering vehicles and heat for cooking (see Chapter 8, Interaction)—with these reactions the objective is to maximize the energy output

- neutralization of acids by alkalis (see Chapter 3, Consequences)

- respiration.

Utilizing exothermic reactions

Combustion reactions use oxygen gas as a reactant and result in the formation of a useful by-product: energy in the form of heat. Such reactions take place in the coal-fired power stations that generate electricity and in internal combustion engines powered by gasoline and diesel.

Questions

1. What factors influence people's choice of cooking fuel?

Around the world many different sources of fuel are used. Each has a different energy density—this is the amount of energy stored in a given mass of a substance. Wood, a traditional means of generating heat energy for cooking and space heating, has a low energy density. It is a relatively inefficient fuel source.

▶ Energy density is a measure of the amount of energy stored in a given amount of substance. It is an important factor when considering the cost of generating energy

The majority of global carbon-based emissions come from the combustion of fossil fuels which are high in carbon content. Combustion of carbon-based fuel adds to the global emission of carbon dioxide, one of the main greenhouse gases.

2. Choose two of the fuels shown in the graph and research their environmental impact.

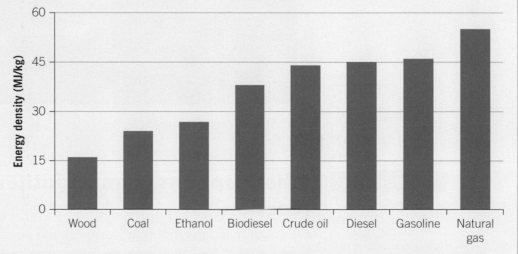

Thinking in context

Is our increasing energy usage sustainable?

Are technological advances in energy production, and the resulting impact on the environment, having an adverse effect?

In groups or as a class discuss the following.

1. The energy needs of the world's population are increasing. What are the reasons for this?

2. The Sun supplies solar energy to our planet; it is limitless and free. Are we utilizing this energy source effectively? What are the challenges in exploiting this source of energy?

3. Some power generation schemes transform their surrounding environment. For example, the Three Gorges Dam is a hydroelectric dam that spans the Yangtze River in China. It is over 2.3 km long. To create its reservoir, over 1 million people had to be relocated. Do the benefits of such projects justify the impact on the environment and its citizens?

▲ Three Gorges Dam

PHYSICAL

What happens in an endothermic reaction?

Chemical reactions that require additional energy from the surroundings are called endothermic reactions. When performing these reactions in the school laboratory, the temperature of the reaction mixture will decrease and the outside of the reaction container, such as a beaker, will feel cold. Energy is transferred from the surroundings into the system.

When barium hydroxide and ammonium chloride are mixed together at room temperature, the temperature of the reaction mixture drops dramatically. This is an example of an endothermic (heat-absorbing) reaction.

 Barium hydroxide and ammonium chloride are placed in a flask sitting on a damp piece of wood. The digital thermometer shows the initial temperature as 21°C. When the two chemicals are mixed, an endothermic reaction takes place and the temperature inside the beaker drops to −23°C. This drop in temperature freezes the water between the flask and wooden block, sticking the two together

 Experiment

Is dissolving a salt in water an endothermic or exothermic reaction?

Ammonium nitrate salt is the product of the reaction between nitric acid and ammonia. Typically it is used in agricultural fertilizers as a source of nitrogen for plants.

⚠ Safety
Wear safety glasses.

Materials

- Two 200 cm³ beakers
- Distilled water
- Thermometer or digital probe
- Magnetic stirrer
- Calcium chloride
- Ammonium nitrate

Method

1. Half fill a clean 200 cm³ beaker with distilled water.

2. Record the temperature of the water.

3. Add 20 g of calcium chloride to the water.

4. Stir the mixture on a magnetic stirrer until all the calcium chloride has dissolved.

5. Record the temperature of the liquid in the beaker.

6. Record your quantitative and qualitative observations.

7. Now repeat with a clean beaker but using 20 g of ammonium nitrate instead of calcium chloride.

Questions

1. How did your observations for the two reactions differ?

2. Identify whether each was endothermic or exothermic.

3. Write a balanced chemical equation for each reaction. Include the state of the reactants and products.

Why are some reactions endothermic and others exothermic?

Chemical reactions involve bond breaking and bond making. To break bonds energy must be adsorbed from the surroundings, and when bonds are formed energy is released into the surroundings.

The balance between these two processes, bond breaking and bond making, determines whether a reaction is endothermic or exothermic overall.

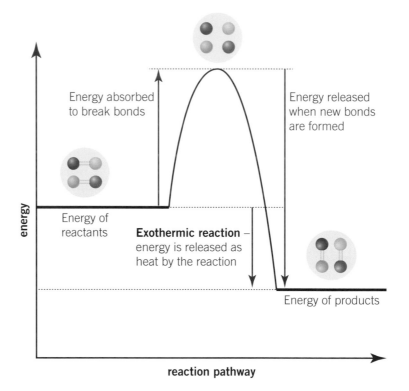

▶ When a reaction occurs and there is an overall decrease in the amount of energy in the system, an exothermic reaction takes place. In the laboratory, an increase in the temperature of the reaction mixture is evidence of an exothermic reaction

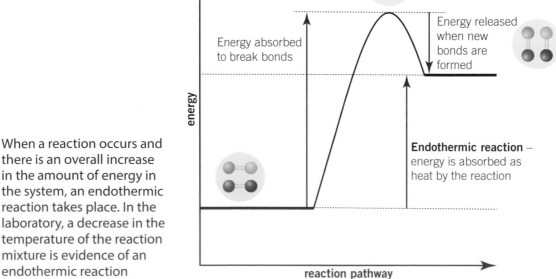

▶ When a reaction occurs and there is an overall increase in the amount of energy in the system, an endothermic reaction takes place. In the laboratory, a decrease in the temperature of the reaction mixture is evidence of an endothermic reaction

Reactions that start without the need for any additional energy input are spontaneous, but other reactions require energy to start them off and, sometimes, to keep them going. To start a reaction, bonds need to be broken and this takes energy. For some reactions, this does not require much energy. For example, to dissolve ammonium nitrate or calcium chloride in water, it is enough to mix the reactants at room temperature.

Chemists work to understand the conditions under which a reaction takes place. This helps them to control reactions, change the rate of the reaction and control the yield of the reaction.

1. Look back at the thermite reaction. Where did the energy come from to start the reaction?

2. In an endothermic or exothermic reaction, what do you observe when measuring the temperature of the reaction mixture with a thermometer?

Why do substances change state?

◀ Chlorine, bromine and iodine are all halogen elements with many properties in common. At room temperature chlorine is a gas, bromine is a dark liquid (although it readily produces a brown vapor) and iodine is a crystalline solid. How do these three states differ from each other?

The arrangement of the molecules determines whether they are a solid, a liquid or a gas.

● A solid has a fixed shape and volume. Its particles are closely packed and can only vibrate about a fixed position.

● If a solid is heated, the particles may gain sufficient energy to break away from their fixed position and move around—the solid becomes a liquid. A liquid also has a fixed volume, but its shape changes—it takes on the shape of its container.

● If a liquid is heated, the particles may gain enough energy to break away from each other and form a gas. In a gas, the particles spread out to fill the container. A gas does not have a fixed volume.

These changes are reversible by cooling.

▶ Ordering of particles in the solid, liquid and gaseous states

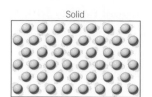
Solid

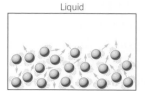
Liquid

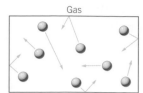
Gas

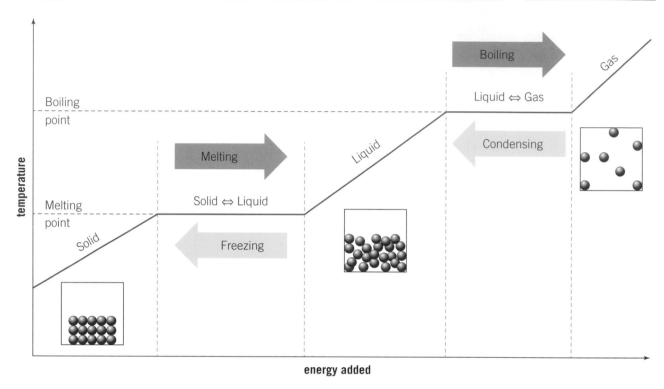

▲ An increase or decrease in the amount of heat energy in a system results in a change of state. Where do you observe changes of state in your daily life?

Most substances can exist as a solid, liquid or gas, but the temperatures at which they transition can vary dramatically. You already know the melting and boiling points of water, but contrast this with the iron formed in the thermite reaction: it has melting and boiling points of 1,540°C and 2,900°C.

It is important to understand that heat and temperature are not the same:

● Heat is a measure of the thermal energy of something. It is measured in joules (J).

● Temperature is a measure of how hot something is. In science, it is measured in degrees Celsius (°C) or kelvin (K).

Measuring temperatures

The Celsius (°C) scale was devised by dividing the range of temperature between the freezing and boiling temperatures of pure water at standard atmospheric conditions into 100 equal parts or degrees.

Absolute zero, −273°C, is the lowest possible temperature. It is considered to be the temperature at which the movement of all particles stops. It is the same for all substances. It led to the development of a new temperature scale—the kelvin (K) scale. It starts at absolute zero with 0 K. Each unit on this scale is called a kelvin and is equal to a degree on the Celsius scale. An advantage of this scale is that there are no negative numbers.

To convert from the Celsius scale to the kelvin scale:

$$temperature\ (K) = temperature\ (°C) + 273$$

1. Convert the following temperatures to kelvin:

   a) 100°C   b) 38°C   c) −48°C

2. Convert the following temperatures to degrees Celsius:

   a) 300 K   b) 273 K   c) 252 K

How much energy does it take to raise the temperature of a substance?

PHYSICAL

When you heat an object, the rise in temperature depends on:

- its mass
- the amount of energy supplied
- its composition.

These quantities are linked by the equation:

heat energy transferred
= mass × specific heat capacity × change in temperature

or

$$q = m\,c\,\Delta T \qquad (\text{the symbol } \Delta \text{ means change in})$$

The specific heat capacity is the amount of heat (in Joules) needed to raise the temperature of 1 kg of a substance by either 1°C or 1 K. The higher the heat capacity, the smaller the rise in temperature of the substance for a given amount of heat energy transferred to the system.

1. It takes 4,200 J to heat 1 kg of water by 1 K (or by 1°C). How much energy does it take to raise the temperature of 1 kg of copper by 1 K?

Substance	Specific heat capacity c (J kg^{-1} K^{-1})
Water	4,200
Copper	390

Worked example: Heat energy transfers

Question 1

Calculate the energy released by the combustion of methanol when the temperature of 30 g of water is raised by 30°C. The specific heat capacity of water is 4,200 J kg^{-1} K^{-1}.

Answer

Convert mass from grams into kilograms.

$$\frac{30g}{1000} = 0.030 \text{ kg}$$

Remember that a temperature change of 30°C is the same as one of 30 K.

$$q = m\,c\,\Delta T$$

$$= 0.030 \times 4200 \times 30$$

$$= 3780 \text{ J} \quad \text{or} \quad 3.78 \text{ kJ}$$

Question 2

0.675 kJ of heat is transferred to 125 g of copper metal. Copper metal has a specific heat capacity of 390 J kg^{-1} K^{-1}. Calculate the change in temperature.

Answer

Convert kilojoules into joules.

$$0.675 \text{ kJ} = 0.675 \times 1000$$

$$= 675 \text{ J}$$

Convert grams into kilograms.

$$\frac{125g}{1000} = 0.125 \text{ kg}$$

Rearrange the equation so the subject of the equation is the change in temperature.

$$q = m\,c\,\Delta T$$

$$\Delta T = \frac{q}{m\,c}$$

$$= \frac{675}{0.125 \times 390}$$

$$= 13.8 \text{ K} \quad \text{or} \quad 13.8°\text{C}$$

1. Convert:

   a) 2.5×10^5 J into kilojoules

   b) 298 K into degrees Celsius

   c) 0.098 kJ into joules

   d) 45°C into kelvin.

2. The temperature of a 4.0 g block of pure aluminium increases from 25°C to 45°C. If the specific heat capacity of aluminium is 900 J kg^{-1} K^{-1}, calculate the amount of energy required for this temperature increase.

3. To what temperature will 250 g of mercury rise from an initial temperature of 10°C, if 5.425×10^3 joules of heat energy is transferred into the system? The specific heat capacity of mercury is 140 J kg^{-1} K^{-1}.

4. A 20 g piece of iron absorbs 1251.2 J of heat energy, and its temperature changes from 25°C to 161°C. Calculate the specific heat capacity of iron.

The reaction between copper(II) sulfate and zinc

Initially 50 cm^3 of 0.20 mol dm^{-3} of copper(II) sulfate is placed in an insulated coffee cup. The initial temperature is 22.0°C. 1.20 g of zinc powder is added to the solution and the change in temperature is monitored. The maximum temperature recorded is 52.4°C.

Mass of copper(II) sulfate solution / g	57.6
Mass of zinc / g	1.20
Initial temperature / °C	22.0
Final temperature / °C	52.4

1. Suggest ways in which heat loss to the surroundings can be minimized.

2. Is this an exothermic or endothermic reaction?

3. What is the change in temperature for this reaction?

4. Calculate the energy given out by this reaction, based on the change in temperature of the solution. Assume that the specific heat capacity of the solution is 4,200 J kg^{-1} K^{-1}.

5. Is your answer likely to be an underestimate or overestimate of the heat given out by this reaction? Explain your reasoning.

Investigating the combustion of alcohols

Different fuels have different energy densities. The combustion of alcohol is an exothermic reaction in which water, carbon dioxide and heat energy are produced. You can compare the energy density of alcohols by using the subsequent heat produced to heat water.

6. Formulate a hypothesis for the relationship between the energy density of alcohols and the length of their carbon chain. Explain this using scientific reasoning. [3]

7. Design an experiment to test your hypothesis using some or all of the materials listed. You should include:

   a) the method; include a drawing of your apparatus

   b) the dependent, independent and control variables

   c) any safety precautions required

   d) a table for recording and analyzing results. [10]

Spirit burner
Alcohol

▲ A spirit burner can be used to burn alcohol

> **Equipment**
>
> - The following alcohols:
>
>   methanol (CH_3OH)
>
>   ethanol (CH_3CH_2OH)
>
>   propan-1-ol ($CH_3CH_2CH_2OH$)
>
>   butan-1-ol ($CH_3CH_2CH_2CH_2OH$)
>
> - Spirit burner
>
> - Electronic balance
>
> - Aluminium drinks can
>
> - Thermometer or temperature probe
>
> - Clamp stand and clamp
>
> - 100 cm^3 measuring cylinder
>
> - Distilled water

a)

b)

▶ Structures of a) methanol, b) ethanol, c) propan-1-ol and d) butan-1-ol

c)

d)

The following table shows the data collected by a chemist when investigating the combustion of a series of alcohols. They may have used a similar method to the one you designed above.

Alcohol	Chemical formula	Initial temperature (°C)	Final temperature (°C)	ΔT (°C)	Mass of fuel consumed (g)	Temperature change per gram of fuel (°C/g)
Methanol	CH_3OH	21.4	54.3		0.71	
		20.9	52.1		0.68	
Ethanol	C_2H_5OH	23.5	67.6		0.79	
		23.1	66.6		0.79	
Propan-1-ol	C_3H_7OH	20.7	66.6		0.68	
		21.1	76.0		0.74	
Butan-1-ol	C_4H_9OH	19.8	57.8		0.47	
		20.2	59.3		0.48	

8. Calculate the change in temperature ΔT (1 decimal place) for each of the trails performed, and record in the results table. [8]

9. Calculate the temperature change per 1 gram of fuel consumed in °C/g, and record in the results table. [8]

10. Of the four alcohols tested, which set of data is the most valid? State the reasons for your choice. [2]

11. a) Which set of data is the least reliable and why? [2]

   b) Explain how you could improve the validity of this data. [2]

How is heat energy distributed around the globe?

Advances in science and technology has enabled us to deepen our understanding of the large-scale effects of different types of energy redistribution. Many of Earth's solid and fluid processes of global significance have the effect of redistributing energy from areas of high concentration towards areas of lower concentration. Examples include:

- the Gulf Stream
- the North Atlantic Drift
- Hadley cells
- hydrothermal venting
- geysers and hot springs
- volcanism.

The scientific community is using some of these systems to track changing global weather patterns. How is energy interconnected within each of these systems?

The Atlantic Meridional Overturning Circulation (AMOC)

More commonly known as the Atlantic conveyor, the AMOC is a part of the global ocean conveyor belt that is responsible for the transport of heat energy around the globe. Heat is defined as the flow of energy resulting from differences in temperature. In simple terms, warm water in the upper region of the Atlantic Ocean flows north from the Earth's equator towards the North Pole.

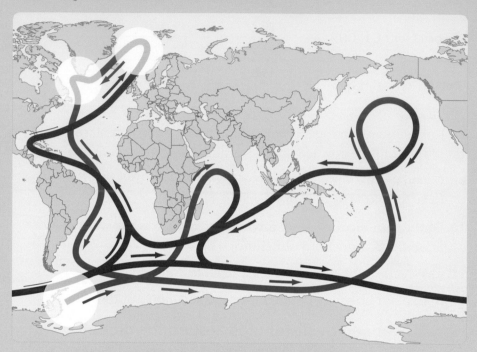

▲ Cold and warm water circulate redistributing energy. Heat is released as the water travels north, so the northern European climate is milder than its geographical position would suggest and the ocean water temperatures are warmer than the equivalent latitudes in the Pacific Ocean. As the water cools, it falls to greater depths and returns south

A number of global longitudinal scientific studies into the effects of man-made global warming on the AMOC have found that the amount of warm water being transported has slowed significantly. The exact reasons for this phenomenon are still unclear. What has been concluded is that changes in the AMOC have resulted in higher sea levels in parts of the northern hemisphere and more severe weather patterns in the affected areas.

The following article was published online in 2017 for the Scripps Institution of Oceanography at https://scripps.ucsd.edu/news/climate-model-suggests-collapse-atlantic-circulation-possible

Read the article and then answer the following questions.

Climate Model Suggests Collapse of Atlantic Circulation is Possible

The primary circulation pattern in the Atlantic is assumed to be stable by most scientists, but new simulation suggests collapse could happen if atmospheric greenhouse gases continue to increase.

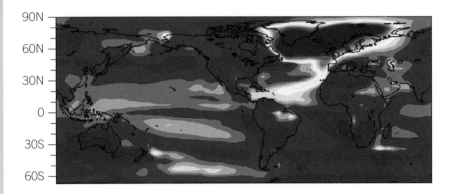

The idea of climate change causing a major ocean circulation pattern in the Atlantic Ocean to collapse with catastrophic effects has been the subject of doomsday thrillers in the movies, but in climate forecasts, it is mostly regarded as an extreme longshot.

Now a new paper based on analysis done at a group of research centers including Scripps Institution of Oceanography at the University of California San Diego shows that climate models may be drastically underestimating that possibility. A bias in most climate models exaggerates the stability of the pattern, called the Atlantic Meridional Overturning Circulation (AMOC), relative to modern climate observations. When researchers removed the bias and re-ran simulations, the result prompted them to predict a collapse of the circulation at some point in the future, setting off large-scale cooling in the North Atlantic. The collapse would stop the AMOC, which delivers warm surface water toward Greenland then sinks as it cools and flows back toward the equator closer to the seafloor.

Wei Liu, a former Scripps postdoctoral researcher now at Yale University, Scripps climate modeler Shang-Ping Xie, and colleagues detail their findings in the paper "Overlooked possibility of a collapsed Atlantic Meridional Overturning Circulation in warming climate" appearing in the journal _Science Advances_ today.

"The significance of our study is to point out a systematic bias in current climate models that hinders a correct climate projection," said Wei. "A bias-corrected model puts the AMOC in a realistic stability regime and predicts a future AMOC collapse with prominent cooling over the northern North Atlantic and neighboring areas. Therefore, our study has enormous implications for regional and global climate change."

In addressing future AMOC change, the paper explores an issue about which there is ongoing debate within the climate science community. The Intergovernmental Panel on Climate Change (IPCC) issues periodic reports that synthesize the latest climate change research. The panel has assumed in its two most recent reports that the AMOC is fundamentally stable and will not collapse, although it might moderately weaken as climate changes. The existence of that bias, though, is widely acknowledged by climate researchers and underscored by recent observations. Some climate modelers have proposed that it is possible that the circulation pattern is prone to collapse, capable of switching between states of multiple equilibria.

The paper uses doubling of atmospheric carbon dioxide concentration as a simple climate change scenario, and relaxes the assumption of AMOC stability. The researchers' simulation showed that the circulation collapses 300 years after the CO_2 concentration doubles its 1990 level of roughly 355 parts per million (ppm) in air.

The effect of the collapse in the model includes a cooling of the northern Atlantic Ocean and a spread of Arctic sea ice. North Atlantic Ocean surface temperatures drop 2.4°C (4.3°F) and surface air temperatures over northwest Europe drop by as much as 7°C (12.6°F). Tropical rain belts in the Atlantic Ocean move farther southward.

The National Science Foundation, the Department of Energy and the Ministry of Science and Technology of the People's Republic of China funded the research. Besides Liu and Xie, report co-authors are Zhengyu Liu and Jiang Zhu of the University of Wisconsin Madison.

"It's a very provocative idea," said Zhengyu Liu. "For me, it's a 180-degree turn because I had been thinking like everyone else."

12. Numerous organizations and scientists are researching the many aspects of global warming. This article presents the point of view that the AMOC will be stopped by a collapse in circulation.

a) Explain the bias that the scientists have named and why they have reached this conclusion about the collapse of the AMOC. [2]

b) What do they predict will happen in the northern North Atlantic Ocean? [1]

c) What problem is identified with the IPCC reports (which synthesize climate change research)? [3]

d) As well as modelling, which other scientific method supports the Chinese scientists' claims? [1]

e) List five possible effects created by the collapse in the model. [5]

5 Conditions

This tidal pool is home to sea stars, sea anemones and other organisms. These organisms that live in habitats between the high and low tide regions must be capable of adapting to the daily change in conditions within the system. During low tide, the animals and plants must adapt to harsh conditions such as periods of exposure to the sun, changing oxygen levels, changes in salinity and predators. Living organisms that live in climates with rapidly changing conditions have learned to adapt, to improve their chances of survival. Global climate conditions are changing more slowly than the daily changes in this tidal pool. What can we learn from changes in conditions in your local habitat?

Coral bleaching observed in the reefs that surround Tioman Island, Malaysia is graphic evidence of the changing conditions in the oceans. As the temperature of the oceans continues to rise, the coral expels the colorful algae which help the coral to survive. What effect will this loss of coral reefs have on the aquatic species that inhabit the coral reef?

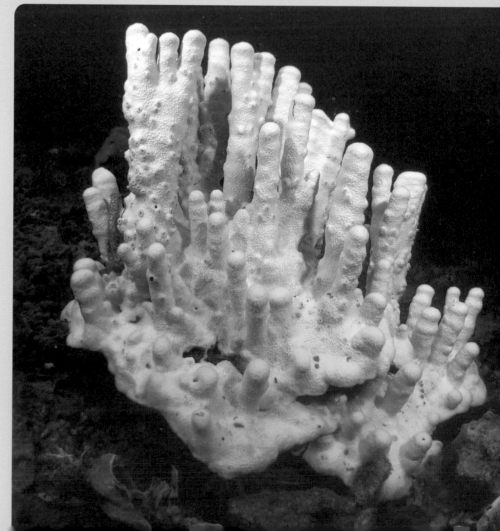

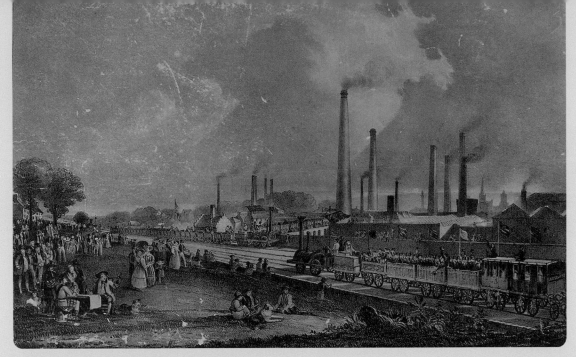

▲ The Industrial Revolution marked the start of the mechanization of industrial processes. Changing physical conditions in manufacturing processes had a profound effect on the daily lives of the general population. Goods and services changed significantly and the positive economic effect was widespread. However, the impact on the environment was also significant, but negative. Scientists research and develop environmentally friendly chemical conditions for synthetic processes, better utilize biotechnology alternatives to traditional chemical reactions and design reaction conditions which are clean and green. Why is Green Chemistry a benefit to the whole planet rather than simply the countries that have adopted this approach to manufacturing?

▲ The Monsanto production process for acetic (or ethanoic) acid involves the reaction between methanol and carbon monoxide. Chemists research and refine the production process, making small changes to the reaction conditions and analyzing the changes in rate of production and yield. The process has an atom economy of 100%, with all of the reactants being converted into the preferred product. Previous methods for the production of acetic acid had an atom economy of 35%. What is the advantage to society and the environment if processes utilize reaction conditions that result in very high atom economies?

Key concept: Systems

Related concept: Conditions

Global context: Scientific and technical innovation

Introduction

When they are performing research, scientists speak of chemical systems. What is a chemical system? When we want to define a chemical system, we need to examine the conditions of the physical and chemical environment inside and outside of the system. In reality, a chemical system is simply a chemical reaction being studied. All matter external to a system is known as the surroundings. The surroundings represent the rest of the Universe.

Chemical systems undergo constant change during a reaction. They are dynamic in nature. We can better understand the changes taking place if we define the type of system being studied. There are three systems that we define in the following way:

▼ Catalysts work to increase the rate of chemical reaction by providing a surface on which specific particles are encouraged to react. Advances in scientific technology now enable scientific research into the actions of catalysts by direct observations of the surface of the catalyst

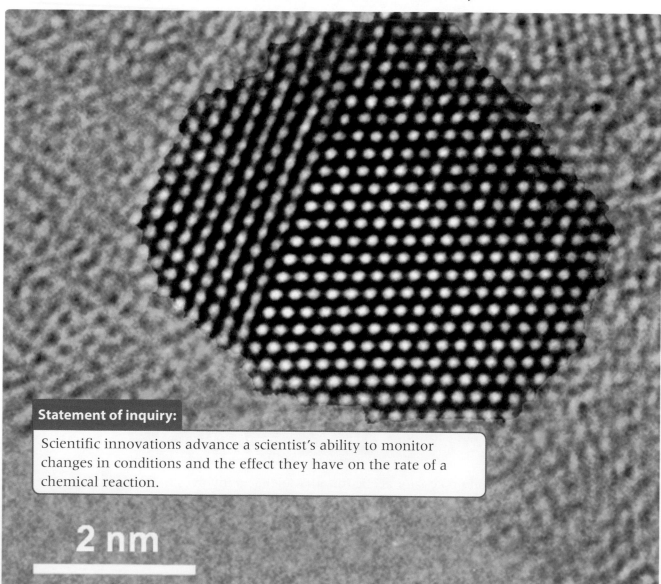

Statement of inquiry:

Scientific innovations advance a scientist's ability to monitor changes in conditions and the effect they have on the rate of a chemical reaction.

2 nm

Type of system	Open system	Closed system	Isolated system
Description	Energy and matter can move in and out of the system freely.	Energy, but not matter, can move in and out of the system.	Neither energy nor matter can move in and out of the system.
Example	Human beings are an example of an open system. We have a dynamic relationship with the environment in which we live. Matter and energy move in and out of our body. We are affected by our environment and in-turn have an effect on the environment.	When you are cooking, a saucepan with a lid is an example of a closed system. Energy continues to enter and leave the system as the food cooks. However, matter cannot escape the saucepan with the lid in place.	An insulated soup thermos is a good example of an isolated system. Both matter and energy cannot enter or leave the system. Your soup keeps hot for a long period of time.

When we understand the characteristics of each type of system it enables us to use the system in more efficient ways. Industrial processes may take place in open, closed and isolated systems. Chemists and chemical engineers working within industry are constantly refining industrial processes to use finite resources more efficiently. Industrial processes that produce fewer by-products and have reduced consequences for the environment are increasingly being recognised as the way of the future.

▶ A pumped-storage power plant takes advantage of the gravitational potential energy of water stored in a high-altitude reservoir. When the water from this top reservoir is released downhill through turbines, its gravitational potential energy is transferred as kinetic energy to generate electric energy. The water is then collected in a bottom reservoir from where it can be pumped back up to the top reservoir. An example of a pumped-storage power plant is in Linthal, Switzerland, which transfers approximately 92 million cubic meters of water between Lake Limmern (1,857 m high) and Lake Mutt (2,474 m). Advances in technology are improving the difference between energy used transferring the water to higher altitudes, and the energy generated as the water is released downhill. This process is a viable clean energy alternative to traditional power supplies

▶ The collision theory explains why some collisions are successful and why some are not

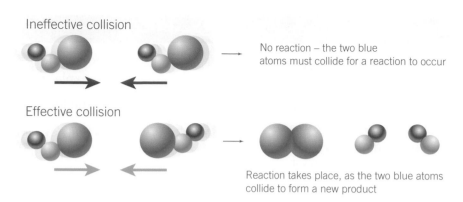

Ineffective collision

No reaction – the two blue atoms must collide for a reaction to occur

Effective collision

Reaction takes place, as the two blue atoms collide to form a new product

1. How can we increase the frequency of these successful collisions in the laboratory?

2. What conditions of a chemical reaction can be altered to increase the rate of a chemical reaction?

PHYSICAL

What is activation energy?

In Chapter 4, Energy we introduced the concept of activation energy, and defined it as the minimum amount of energy colliding particles need for a collision to be successful, and for a reaction to occur. The amount of activation energy required for a reaction to proceed determines how spontaneous that reaction is for a given set of conditions.

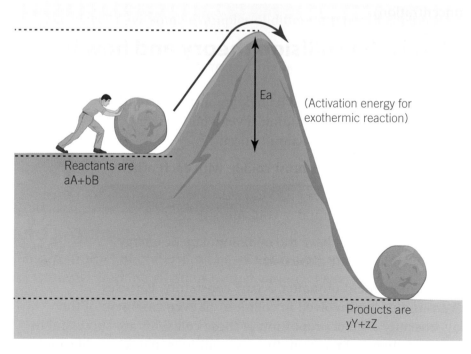

E_a

(Activation energy for exothermic reaction)

Reactants are
aA+bB

Products are
yY+zZ

▲ Depending on the reaction conditions, not all reactants will have sufficient energy to overcome the activation energy barrier and be involved in the chemical reaction. Can you name some simple reactions that will occur more readily when the temperature is increased?

Gathering and organizing relevant information

Learning is a lifelong journey. Sometimes we learn in a more structured, formal environment like a school, university or college. Many times, we learn through experience as we pass through life. Information can be organized to help establish relationships between facts and develop our understanding.

Here is some additional information on chemical kinetics:

- The activation energy of a reaction is the minimum energy barrier reactant particles must overcome in order to react.

- The SI unit for activation energy is kJ/mol or kJ mol^{-1}.

- The activation energy of a reaction is determined experimentally using the Arrhenius equation.

- The activation energy of a reaction is temperature dependent.

What resources would cover a more in-depth discussion of these points? What learning skills would you utilize to enable you to increase your understanding of chemical kinetics?

How is the rate of a chemical reaction measured?

PHYSICAL

Often, the reactants involved in industrial chemical processes are natural resources, which are finite. It is therefore important to use these resources effectively. One area of scientific research focuses on redesigning chemical processes to minimize waste.

When considering how a chemical reaction will proceed from the start of the reaction to completion, there are a number of things we need to consider:

- What is occurring when the reactants collide?

- What proportions of the starting materials have successfully reacted to form new products?

- How quickly has the reaction occurred?

- What is happening to the distribution of energy during the reaction?

- Does the reaction require energy to be transferred from the surroundings into the system? If so, this is called an endothermic reaction.

- Does the reaction produce energy to be transferred into the surroundings from the system? If so, this is called an exothermic reaction.

- Is the reaction spontaneous or non-spontaneous for a given set of conditions?

Observations can provide us with important information about reactions that are taking place. What are some observations you can make to infer a chemical reaction is taking place?

Observations

The characteristic yellow yolk and the colorless albumen make it easy to recognise a chicken egg.

When you start to cook an egg in a frying pan, what changes in appearance do you observe?

These observations are qualitative observations: they are non-numerical descriptions of what you are seeing. Were all of your observations qualitative?

Quantitative data or numerical data is collected and analyzed to monitor the progress or rate of a chemical reaction. Sometimes you will hear the term "empirical data" being used. This is any type of data that is collected under experimental conditions in a laboratory.

The rate of reaction is defined as the change in amount of either reactants or products per unit of time:

$$\text{Rate of reaction} = \frac{\text{change in amount of reactant or product}}{\text{time taken}}$$

The progression of a reaction can be graphically represented, allowing you to make calculations of the initial rate of reaction. The units for the rate of reaction are dependent on the measurements being taken:

- If a gas is being produced in the reaction, the rate of reaction is measured in $cm^3\ s^{-1}$.

▼ The initial rate of reaction is represented by a tangent to the curve at 0 seconds. When you examine the curve over the duration of the reaction, the gradient of the tangent slowly decreases until the gradient equals zero. How does the shape of the curve in chemical kinetic studies help us to understand what is happening as the reaction proceeds?

- Measuring the initial rate of a reaction between two aqueous solutions requires monitoring of the concentration of one or both reactants; the units of measurement would be $mol\ dm^{-3}\ s^{-1}$.

- The instantaneous or initial rate of a reaction can be determined by calculating the gradient of a tangent drawn to the curve at t=0 s.

Chemical reactions are at the centre of many useful processes in a wide variety of industries, so it is important not only to determine the rate at which they occur, but also to what extent they can be controlled. The

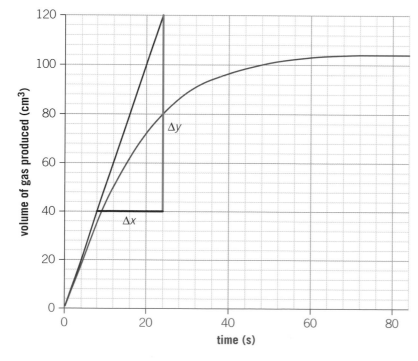

conversion of raw materials into useful products is the focus of synthesis reactions. The plastics industry, pharmaceutical laboratories, food technology, energy suppliers and chemical industries such as the production of ammonia, sulfuric acid and fertilizer, all benefit from the work of chemists and chemical engineers controlling the rate of the chemical reactions.

How can changes in the mass or volume of a gas be used to monitor the rate of a reaction?

PHYSICAL

In all chemical reactions mass is conserved. Atoms in the reactants are rearranged to form the products.

Some chemical reactions result in the production of a gas. An example is the reaction between calcium carbonate, $CaCO_3$, and hydrochloric acid:

$$CaCO_3(s) + 2HCl(aq) \rightarrow CaCl_2(aq) + CO_2(g) + H_2O(l)$$

Looking at the reaction, which of the products could we monitor to measure the rate of reaction? If this is an open system, the gas is free to leave the system; the amount of matter in the system has decreased.

In general, we can monitor the rate of a reaction by looking at the production of gas, in this case carbon dioxide. One technique commonly used in a school laboratory is to monitor the loss of mass of the reaction mixture. If the gas produced is allowed to escape the system, a loss of mass will occur. This loss of mass can be measured accurately using an electronic balance. The quantitative data collected in this way can be graphed, and the shape of the graph can be analyzed to determine how the reaction progresses over time.

Data-based question: Determining the rate of reaction

20 cm³ of 1.0 mol dm⁻³ hydrochloric acid was added to 5.0 g of calcium carbonate chips in a conical flask. The formation of gas over time was measured and recorded in the table below:

Time (min)	1	2	3	4	5	6	7	8	9	10
Volume of CO₂ (cm³)	14	26	36	43	50	59	64	70	74	78

1. Plot the volume of carbon dioxide gas evolved $V(CO_2)$ against time t. You can do this by hand or by using graphing software. Remember that a smooth curve is required to represent the relationship between the data points.

2. What conclusions can you draw about the progress of the reaction based on the shape of the curve?

▲ It is possible to monitor gas production by measuring loss of mass

3. Draw a tangent to the curve at t=0 min and calculate the gradient of the tangent. What are the units for this rate?

4. From the shape of the curve, estimate the time at which the reaction has finished.

Research skills

Presenting information and data using models and mathematical relationships

Research methods of monitoring the production of a gas in a chemical reaction that are safe to use in a school laboratory.

When critically reviewing the information available, consider how you will present this information to the intended audience:

- How will you communicate the experimental methods most effectively?

- What are the similarities and differences in the chosen methods?

- If these methods are applied to the reaction between calcium carbonate and hydrochloric acid, how would you graphically represent the empirical data collected?

- What are the differences in the shape of the graphs?

PHYSICAL

How can we monitor pH change in acid base neutralization reactions?

When an acid reacts with a base, a neutralization reaction takes place; for example, hydrochloric acid, a strong acid, can be neutralized by sodium hydroxide, a strong base:

$$HCl(aq) + NaOH(aq) \rightarrow NaCl(aq) + H_2O(l)$$

The products of a neutralization reaction are a salt and water.

When a strong acid is titrated with a strong base, the initial pH of the solution will be approximately 2; as more base solution is added, the pH gradually increases. What will the pH be when all of the acid has reacted?

A neutralization reaction is an open system where both energy and matter can be transferred into and out of the system. For this chemical reaction, energy is transferred out of the system, as acid–base neutralization is an exothermic reaction.

▲ Changes in pH can be monitored using a pH probe

How does changing the temperature of the reaction mixture affect the rate of reaction?

PHYSICAL

The process of cooking involves many different chemical reactions. From experience, you know that the rate of cooking a meal is dependent on temperature. For example, cooking in an oven will take less time if the temperature is 230°C rather than 180°C.

Like food, the rate of chemical reactions in a laboratory is dependent on the amount of heat energy applied, with some chemical reactions being more sensitive to changes in the temperature of the reaction mixture.

The activation energy of a chemical reaction is specific for that reaction. Reactions which require little additional transfer of heat energy from the surroundings into the system to proceed are described as having a low activation energy. If a large amount of heat energy needs to be transferred from the surroundings into the system to initiate the chemical reaction, the reaction is endothermic and has a large activation energy.

We now understand that for a reaction to occur the reacting particles need to collide with sufficient energy to overcome the activation energy barrier. With this in mind, what effect will increasing the temperature of a reaction mixture have on the rate of a reaction?

▲ Chinese cuisine is cooked in a wok over very high heat, and only takes a few minutes to prepare

Not all students running around in the playground possess the same amount of energy

Unpicking a metaphor

Increasing the temperature of the reaction mixture involves a transfer of heat energy to the reacting particles; in other words, energy from the surroundings enters the system. We can understand this better by imagining the scene at break time in the playground of a primary school: the students run around with great energy.

The primary school students are a metaphor for the reacting particles within a system. Through discussion with your peers, develop an explanation of how increasing energy or temperature will increase the rate of a chemical reaction. Your explanation should include the following terms:

- Thermal energy
- Kinetic energy
- Frequency of collision
- Activation energy

▶ In an exothermic reaction, for the reaction to proceed, reacting particles need to overcome the activation energy barrier. If your science class was set the challenge to walk over a nearby hill, would all of your friends have the same amount of energy to complete the task?

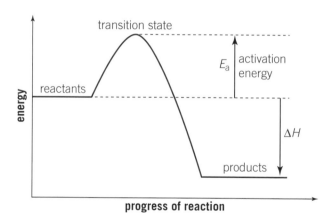

A proportion of the reacting particles in any reaction will have sufficient energy to overcome the activation energy barrier and transform from reactants into products. Increasing the temperature of the reaction mixture will increase the average kinetic energy of the particles and in turn increase the frequency of successful collisions.

▶ As temperature increases, so does the proportion of reactants with sufficient energy to overcome the activation energy barrier. The area under the curve, above the activation energy, represents the number of reacting molecules. Which temperature has the highest number of reacting molecules? Which temperature will have the highest yield?

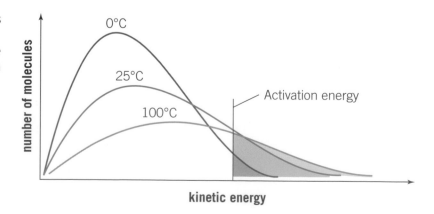

The reaction conditions can have a direct affect on the rate of a chemical reaction.

In industrial processes, increasing the temperature of the reaction mixture is costly, and the decision of whether to invest is taken by considering whether the benefits outweigh the costs.

 Experiment

How does changing the temperature conditions affect the rate of a chemical reaction?

 Safety

- Wear safety glasses at all times

- If available, perform this demonstration in the fume hood, otherwise it should be performed in a well-ventilated area of the laboratory. The product sulfur dioxide can cause asthma attacks.

Materials

- Water baths at temperatures of 25°C, 35°C, 45°C, 55°C and 65°C

- 2.0 mol dm^{-3} sodium thiosulfate solution

- 1.0 mol dm^{-3} hydrochloric acid solution

- 50 cm^{-3} measuring cylinder

- 250 cm^{-3} conical flask

- Stopwatch

Method

1. Measure 10 cm^{-3} of 2.0 mol dm^{-3} sodium thiosulfate solution using a measuring cylinder and transfer to a 250 cm^{-3} conical flask.

2. Add 40 cm^{-3} of distilled water to the conical flask and swirl the solution to mix well.

3. Prepare a white sheet of paper on which you have drawn a black cross, X.

4. Carefully measure 5 cm^{-3} of 1.0 mol dm^{-3} hydrochloric acid using a 10 cm^{-3} measuring cylinder.

5. Place the conical flask into the 25°C water bath. Check the temperature of the sodium thiosulfate solution with a thermometer. When the solution has reached the

same temperature as the water bath, the experiment is ready to begin.

6. Remove the conical flask from the water bath and place it on top of the black cross on the white sheet of paper.

7. Add hydrochloric acid to the conical flask and immediately swirl the mixture and start your stopwatch.

8. Observe the black cross by looking down into the conical flask through the reaction mixture. When you decide you can no longer see the black cross, stop the stopwatch.

9. Repeat the procedure with samples of sodium thiosulfate that have been warmed to 35°C, 45°C, 55°C and 65°C.

Data analysis and questions

1. Draw up a data table using the following headings:

Temperature of the reaction mixture in the conical flask (°C)	Time taken for the black cross to disappear (s)	Rate of reaction (s^{-1})

2. Calculate the value of 1/ time and record in your data table for each of the individual trials.

3. Plot a graph of rate versus temperature and draw a smooth curve connecting the points.

4. Identify any anomalous data and ignore these data points.

5. Analyze the graph and discuss what effect, if any, an increase in the temperature conditions has on the rate of the reaction.

6. Use your knowledge of the collision theory to explain the effect changing temperature conditions has on the rate of reaction.

PHYSICAL

How does changing the concentration of a reactant affect the rate of reaction?

Increasing the number of moles of the limiting reagent in a reaction, while maintaining the temperature and volume of aqueous solutions, will increase the rate of the chemical reaction (see Chapter 9, Models). A higher concentration of reactants equates to more reacting particles in a given area, which in turn leads to a higher frequency of a successful collision.

▶ An increase in the concentration of hydrochloric acid results in an increase in the number of particles colliding with the surface of the magnesium metal. A chemical reaction is occurring. What observation could be made, that would indicate an increase in the rate of the chemical reaction?

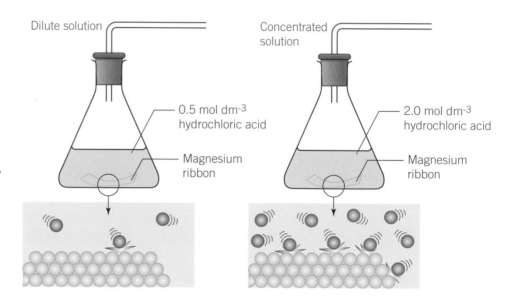

Dilute solution

0.5 mol dm⁻³ hydrochloric acid

Magnesium ribbon

Concentrated solution

2.0 mol dm⁻³ hydrochloric acid

Magnesium ribbon

PHYSICAL

How does a change in the surface area of a reactant affect the rate of the reaction?

Picture the layers of an onion. If you wanted to fry a whole onion in oil, it would be a difficult and slow process as only a small surface area of the onion is exposed to the hot oil. How could you fry the onion faster at the same temperature?

  Experiment

How does changing the concentration of a reactant affect the rate of a chemical reaction?

⚠ **Safety**

- Wear safety glasses at all times
- Hydrochloric acid is a corrosive liquid.

Materials

- 2.0 mol dm⁻³ sodium thiosulfate solution
- 1.0 mol dm⁻³ hydrochloric acid solution
- 50 cm⁻³ measuring cylinder

- 250 cm^{-3} conical flask
- Stopwatch

Method

1. Measure 50 cm^{-3} of 2.0 mol dm^{-3} sodium thiosulfate solution using a measuring cylinder and transfer to a 250 cm^{-3} conical flask.

2. Prepare a white sheet of paper on which you have drawn a black cross, X.

3. Carefully measure 5 cm^{-3} of 1.0 mol dm^{-3} hydrochloric acid using a 10 cm^{-3} measuring cylinder.

4. Place the conical flask on top of the black cross on the white sheet of paper.

5. Add hydrochloric acid to the conical flask and immediately swirl the mixture and start your stopwatch.

6. Observe the black cross by looking down into the conical flask through the reaction mixture. When you decide you can no longer see the black cross, stop the stopwatch.

7. Create different dilutions of the sodium thiosulfate solution by successfully increasing the volume of water and decreasing the volume of sodium thiosulfate. The dilutions are found in the data table below.

8. Repeat steps 3–6 for each different concentration.

Data analysis and questions

1. Record your data in a table with the following headings:

Volume of water (cm³)	Volume of sodium thiosulfate (cm³)	Time taken for the black cross to disappear (s)	Rate of reaction (s⁻¹)	Concentration of thiosulfate solution (mol dm⁻³)
0	50			2.0
10	40			1.6
20	30			1.2
30	20			0.8
40	10			0.4

2. Calculate the value of 1/ time and record in your data table for each of the individual trials.

3. Plot a graph of rate versus concentration and draw a smooth curve connecting the points.

4. Suggest why this analytical method of monitoring the rate of the reaction is appropriate for this chemical reaction.

5. Identify the independent and dependent variables.

6. Discuss which variables are being controlled and offer scientific reasons why this is important for this type of reaction.

7. Use your knowledge of the collision theory to explain the effect concentration has on the rate of reaction.

8. Explain why the rate of reaction changes over time. Your answer should include references to the collision theory.

▶ The process of digestion begins in the mouth, where chewing increases the surface area of the food exposed to biological catalysts called enzymes. The transfer of nutrients from food into the body is essential for life

By separating the layers from each other, you effectively increase the surface area of the onion exposed to the cooking oil. This also happens when you chop the onion.

Increasing the surface area of a chemical reactant increases the number of particles present in the reaction mixture. You have not changed the actual amount (in mol) or mass of the reactant. You simply have given more of the particles the chance to undergo a collision and react if they have sufficient energy. It is similar to increasing the concentration of the reactant except overall the number of moles of reactant remains unchanged.

Increasing the surface area of a reactant is one of the simplest and cost effective ways to increase the rate of a chemical reaction in industry. Mechanical grinding on small and large scales produces finely divided solids, so the amount of finite resources used is constant, but the rate of the reaction can increase. Preparing reactants in this way is cost effective, as no extra chemicals are required, and low waste, as it does not generate any unwanted or hazardous products. Can you find some examples of where industry increases the rate of reaction by crushing and pulverising the reactants?

Increasing the surface area of some reactants can sometimes result in explosive reactions. In 2008, many people were killed as a result of an explosion caused by combustible sugar dust. Companies such as sugar and flour refineries are careful to extract and filter the air in these factories to prevent explosions. What other industries use substances that can be explosive when they are present as a finely divided powder?

Experiment

How flammable is cornstarch?

 Safety

- Wear safety glasses at all times.

Materials

- Powdered cornstarch
- Bunsen burner
- Plastic dropping pipette

Method

1. Place a small quantity of cornstarch on a teaspoon. Press it onto the spoon so that it is compressed.

2. Place the spoon onto a heat proof mat and light a Bunsen burner. Carefully direct the flame onto the compacted cornstarch. Make your observations.

3. Fill a plastic dropping pipette with cornstarch. Tap the tip of the pipette onto the benchtop to ensure the cornstarch is at the tip of the pipette.

4. Hold the pipette about 30 cm from the Bunsen burner flame, dim the lights of the classroom, and in one quick motion, carefully drop the cornstarch into the open flame.

Questions

1. What differences were there in the observations you made of the two different combustion reactions?

2. Why was there a difference in the rate of reaction? Support your statement with scientific reasoning.

 Experiment

How does changing the surface area of a reactant affect the rate of a chemical reaction?

 Safety

- Wear safety glasses at all times
- Hydrochloric acid is a corrosive liquid.

Materials

- Calcium carbonate (marble) chips in small and large sizes
- 2.0 mol dm^{-3} hydrochloric acid solution
- Top-pan electronic balance which can read to 0.01 grams
- 100 cm^3 conical flask
- Cotton wool balls
- Conical flask
- 50 cm^3 measuring cylinder
- Stopwatch

Method

1. Using a measuring cylinder, measure 30 cm^3 of 2.0 mol dm^{-3} hydrochloric acid solution and transfer to a 100 cm^3 conical flask.

2. Using the electronic balance, weigh 20 g of large calcium carbonate chips.

3. Place the conical flask on the weighing pan of the electronic balance and press the TARE key to zero the display.

4. Quickly transfer the calcium carbonate chips into the conical flask, place a cotton wool ball into the mouth of the conical flask and start the stopwatch.

5. Record the initial mass of the reaction mixture as the mass at t = 0 seconds.

6. Every 30 seconds, take a reading of the mass and record this in a data table.

7. Continue taking readings until the loss of mass stabilizes and becomes constant.

8. Prepare a fresh solution of hydrochloric acid in a clean conical flask.

9. Using the electronic balance, weigh 20 g of small calcium carbonate chips.

10. Repeat steps 3–7.

Data analysis and questions

1. Record your data in a table with the following headings:

Time (minutes)	Mass reading (g)	Loss of mass of large marble chips (g)	Mass reading (g)	Loss of mass of small marble chips (g)

2. Prepare a graph with loss of mass of the marble chip on the y-axis and time on the x-axis.

3. Plot the two sets of data on the one set of axes to enable you to make a comparison of the data.

4. Draw smooth curves to match the trends of the data and label the curves.

5. Identify and write a balanced chemical equation for the reaction between calcium carbonate and hydrochloric acid.

6. Suggest why this analytical method of monitoring the rate of the reaction is appropriate for this chemical reaction.

7. Identify the independent and dependent variables.

8. Discuss which variables are being controlled and offer scientific reasons why this is important for this type of reaction.

9. Use your knowledge of the collision theory to explain the effect surface area has on the rate of reaction.

10. Explain why the rate of reaction for both investigations changes over time.

   ● Your answer should include references to the collision theory.

   ● Your answer should include example calculations of the rate of reaction at different times.

11. Give reasons why the final loss of mass in each investigation should be identical.

What is the role of a catalyst in a chemical reaction?

The ozone layer is a naturally occurring barrier found in the stratosphere, 10 to 50 km above the Earth's surface. It plays an important role in protecting our planet from harmful ultraviolet (UV) radiation from the sun, which would otherwise damage cells in both plants and animals.

When UV radiation reaches the upper atmosphere, ozone molecules absorb this form of energy and are split:

$$O_3(g) \xrightarrow{\text{UV radiation}} O_2(g) + O(g)$$

Ozone is then reformed and heat is released in the process. There is a net energy transfer from UV radiation to heat energy:

$$O_2(g) + O(g) \rightarrow O_3(g) + \text{energy}$$

In the early 1970s, researchers discovered that a class of compounds used in air conditioning systems and refrigerators as cooling agents, known as chlorofluorocarbons (CFCs), can also react with ozone in the stratosphere. Although CFCs were considered to be harmless, when exposed to UV radiation in the stratosphere they decompose to produce chlorine. Chlorine can then act as a catalyst to split ozone into molecular oxygen. As catalysts are not used up in chemical reactions, a small amount of chlorine can cause a large amount of ozone to be split, in a process generally known as ozone depletion.

The depletion of the ozone layer is an example of the impact human activity has on the global environment. This results in an increase in UV radiation reaching the Earth's surface, causing an increase in skin cancers and eye diseases.

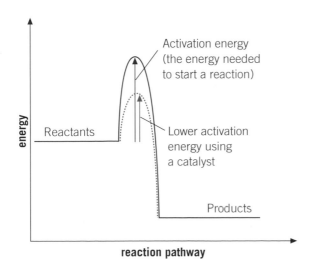

◀ The addition of a catalyst to a reaction system provides an alternative reaction pathway of lower activation energy. This results in a higher proportion of reactants with sufficient energy to overcome the activation energy barrier and transform into products

Data-based question: Stratospheric ozone data

The World Meteorological Organization-Global Atmosphere Watch (WMO-GAW) monitors global air quality and analyzes the patterns in the composition of atmospheric gases and particulate matter.

The graphs below represent a longitudinal study of the concentration of carbon dioxide, methane and nitrous oxide, expressed in parts per million (ppm) or parts per billion (ppb), in gas samples at Cape Grim in Tasmania (Australia) over the last 40 years.

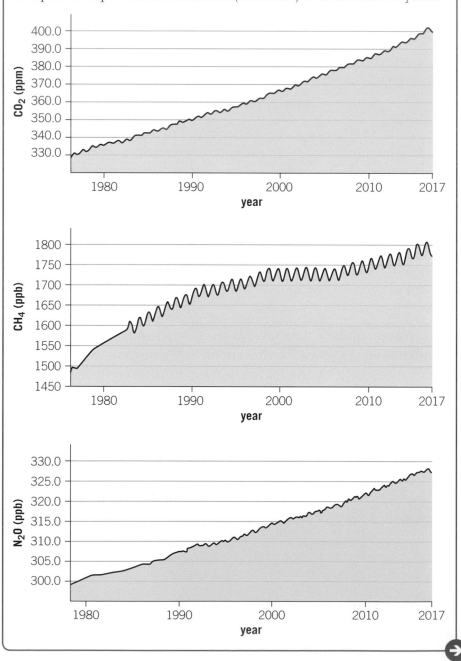

1. Calculate the percentage increase in each of these gases for the time periods 1980–1989, 1990–1999, 2000–2009 and 2010–2017.

2. Compare the percent increase for each time period and discuss the change in concentration of these gases, with time. Is the change constant?

3. Compare the concentration of carbon dioxide recorded in 2017 to the concentration of methane and nitrous oxide in the same year. By what factor do these concentrations differ?

4. What is the main source of nitrous oxide gas in the Earth's atmosphere? State and justify if there is a link between global development and the concentration of nitrous oxide gas.

The graph below shows global concentration of carbon dioxide from 2013 to the start of 2017. The dashed red line shows the mean value of carbon dioxide concentration per month. The blue line represents the same, but takes the seasonal cycles of carbon dioxide concentration into account, and is corrected as such.

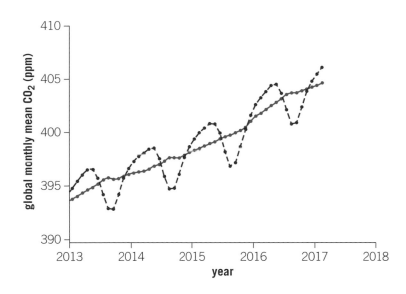

Source of data: *Ed Dlugokencky and Pieter Tans, NOAA/ESRL (www.esrl.noaa.gov/gmd/ccgg/trends/)*

5. Compare the concentration of carbon dioxide in Tasmania and globally over the period 2013–2017. Why are we able to compare this data and make comparisons even though the data originates from different parts of the globe?

Summative assessment

Statement of inquiry:

Scientific innovations advance a scientist's ability to monitor changes in conditions and the effect they have on the rate of a chemical reaction.

Introduction

In this summative assessment, we will look at the impact changing conditions have on the rate of a chemical reaction. We will then consider how the management of finite resources in chemical systems can have an impact on the global environments.

 The effect of change in concentration on the rate of reaction

1. In terms of energy, explain why only a certain proportion of reacting particles in a chemical system will be transformed from reactants into products, at a given temperature. [2]

2. Applying your understanding of chemical kinetics, suggest reasons why science needs to find ways to increase the rate of a chemical reaction. [3]

3. The reaction between calcium carbonate, a metal carbonate, and hydrochloric acid, a strong acid, is a vigorous reaction that produces a gas, carbon dioxide.

   a) Write an equation for the reaction between calcium carbonate and hydrochloric acid. [3]

A series of simple experiments were performed and the data collected was plotted on a graph:

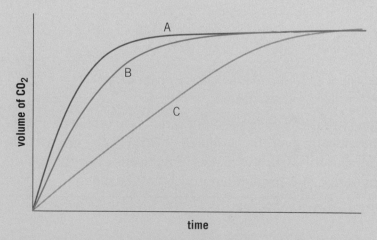

   b) The three curves are the result of different concentrations of hydrochloric acid. Which curve represents the highest concentration and which represents the lowest? Justify your answer using scientific reasoning. [4]

c) All of the curves finish at the same point on the graph. What does this tell us about the reaction? [2]

d) Predict the limiting reagent in this reaction. [1]

The collision theory describes three conditions that need to be met by particles, in order for them to react successfully. When these conditions are met, a reaction will proceed at a rate specific for a given set of conditions. We have learnt that you can change the rate of a reaction by altering the temperature, concentration and surface area of reactants or by adding a catalyst.

The thiosulfate ion reacts with hydrogen ions supplied by hydrochloric acid, a strong acid. Elemental sulfur forms a suspension and is the reason for a color change in the reaction mixture:

$$S_2O_3^{2-} (aq) + 2H^+ (aq) \rightarrow H_2O(l) + SO_2(g) + S(s)$$

4. Design an experiment between 1.0 mol dm^{-3} sodium thiosulfate and 1.0 mol dm^{-3} hydrochloric acid to achieve a color change from colorless to yellow in 60 seconds exactly. You are permitted to change any variable to achieve this outcome. You could use the following points for guidance. [10]

- Think of a question you will test in your investigation.

- Formulate a testable hypothesis supported by scientific reasoning and theoretical knowledge.

- Decide and explain what variables you are manipulating: identify the dependent and independent variables. What variables will you control?

- Your method must enable sufficiently relevant data to be collected. All apparatus used should be included in the method.

- The quantitative and qualitative data you collect should be organized and presented appropriately.

- The subsequent analysis of the data will enable you to reach a conclusion that the data can support.

Safety

- Acids are corrosive and should be handled safely. Avoid contact with the skin.

- Safety glasses must be worn at all times.

Equipment

- 1.0 mol dm^{-3} sodium thiosulfate

- 1.0 mol dm^{-3} hydrochloric acid

- 250 cm^3 conical flask

- 50 cm^3 and 10 cm^3 measuring cylinder

- Stopwatch

Activation energy of a simple chemical reaction

Some students performed the reaction between the thiosulfate ion and hydrogen ions at a range of temperatures. They then recorded their experimental data, using the following table.

Temperature (°C)	Temperature (K)	Time (s)	$1/T$ (s⁻¹)
10		250	
21		87.3	
37		31.2	
42		25.9	

5. Why it is important that you carefully consider how you will present your quantitative data? [2]

6. Copy the data table and calculate the temperature in kelvin (K). [4]

7. Construct a temperature (K) versus time (s) graph and draw the appropriate smooth curve. [4]

8. Identify and describe the relationship between temperature and time. [1]

9. Calculate inverse time (s⁻¹) and record the process data to three significant figures. [4]

10. Construct a temperature (K) versus $1/T$ (s⁻¹) graph. State the relationship illustrated by this form of graph. [3]

11. Explain why it is useful for different graphs to be used to represent experimental data. [3]

12. Having examined the data collected, comment on the accuracy of the data given the shape of the curve. Are there any data points that you would consider to be anomalous? Anomalous data is that which does not follow the general trend. [2]

13. Evaluate and make a conclusion about the effect of concentration on the rate of reaction and support this conclusion with scientific reasoning. [3]

 The importance of catalysts

The article here was written by XiaoZhi Lim and published in the scientific journal Nature, in 2016 (http://www.nature.com/news/the-new-breed-of-cutting-edge-catalysts-1.20538). Read the following extract carefully and then complete the tasks that follow.

The new breed of cutting-edge catalysts

Catalysts are used in some 90% of processes in the chemical industry, and are essential for the production of fuels, plastics, drugs and fertilizers. At least 15 Nobel prizes have been awarded for work on catalysis. And thousands of chemists around the world are continually improving the catalysts they have and striving to invent new ones.

That work is partly driven by an interest in sustainability. The aim of catalysis is to direct reactions along precisely defined pathways so that chemists can skip reaction steps, reduce waste, minimize energy use and do more with less. And with growing concerns about climate change and the environment, sustainability has become increasingly important. Catalysis is a key principle of 'green chemistry': an industry-wide effort to prevent pollution before it happens.

Catalysts are also seen as the key to unlocking energy sources that are much more inert and difficult to use than coal, oil or gas, but much cleaner. Catalysis can make it more economically feasible to split water into oxygen and hydrogen fuel, or can open up new ways to use raw materials such as biomass or carbon dioxide. "These are feedstocks that are ripe for advances in catalysis," says Melanie Sanford, a chemist at the University of Michigan in Ann Arbor.

These challenges have led to an explosion in catalyst innovation, with the annual number of publications on the subject tripling in the past decade. Many groups are coming up with new small-molecule complexes or are chemically tailoring biological enzymes in search of radically new catalytic activity. Others are pursuing advances in nanotechnology, which allow them to engineer the action of solid catalysts at the atomic scale. Still others are experimenting with catalysts that are activated by light, or that incorporate the DNA double helix. And everyone in the field is trying to streamline the search for better catalysts with modern computational modelling tools.

The pace of innovation is such that even the experts are struggling to keep up, says Scott, who leads the US Department of Energy's efforts to develop benchmarks for the new catalysts' performance. "We need to make sure we are advancing the science that's most efficient," she says.

And the scope of catalysis is increasing rapidly. "Twenty years ago," says John Hartwig, a chemist at the University of California, Berkeley, "catalysis to make molecules that were complex did not exist." Anyone who wanted to modify a large complicated structure would have to tear it down and build it back up, says Sanford. But now, chemists can often edit parts of a molecule precisely. "It's incredibly enabling," she says.

Cut-price catalysts

Using a catalyst is like bulldozing a shortcut between reactants A and product B, bypassing convoluted chemical pathways that might otherwise take forever. Using a really good catalyst is like building a multi-lane superhighway. And some of the best are the "homogeneous" catalysts: free-floating molecules that are mixed in with the reactants.

Industrial catalysts in this category most often consist of a metal ion that does the hard work of making or breaking chemical bonds, surrounded by "ligands": connected groups, often carbon-based, that control the reactants' access to the ion. Much of the research in this field comes down to tailoring these ligands to produce a catalyst that performs only a desired reaction.

Unfortunately, many of the successes so far have come through the use of scarce and expensive metals such as palladium, platinum, ruthenium and iridium. Today, chemists are increasingly striving to build catalysts around cheaper, "Earth-abundant" elements such as iron, nickel or copper – or to do without metals altogether.

14. How does the process of catalysis impact on sustainability? [2]

15. Discuss the positive implications of the use of catalysts in energy production. [2]

16. Evaluate and explain how advances in computing power are assisting research in the field of catalysis. [2]

17. Many catalysts are made of expensive, scarce metals such as palladium, platinum and iridium. Scientists are attempting to address this issue by researching alternative catalysts for industrial processes. Describe the benefits of the use of cheaper, more abundant metals such as iron, nickel or copper as an alternative to more expensive, rare metals. [2]

6 Form

◀ The Japanese-style garden or Zen garden has a very specific form that is easily recognized. Originating from Buddhist temples, the gardens were designed by monks to provide a place for meditation and reflection. Common features of a Japanese garden include raked white stones, carefully chosen rocks, manicured groundcover plants such as moss and shrubs, and carefully cultivated trees. How does form influence the aesthetics of objects in nature?

▼ Species of plants and animals are classified based on their form. Koi carp are often referred to as goldfish, but they are the product of selective breeding of common carp in Japan in the early 18th century, whereas goldfish were selectively bred in China over 1,000 years ago. How do the classification systems used for plants and animals compare to the systems used to classify elements and compounds?

◀ Salicylic acid, more commonly known as aspirin, is a widely used medication for the relief of inflammation, pain and fever. A derivative of a natural compound found in the bark of willow trees, the compound has a characteristic crystalline form which can be observed under polarized light. How would you describe the characteristics of this crystal of salicylic acid?

▼ Graphene is composed of a single 2D layer of carbon atoms covalently bonded to one another. Graphene has far greater electrical conductivity than copper, and is hundreds of times stronger than steel at a fraction of the mass. Russian scientists Andre Geim and Konstantin Novoselov first isolated graphene from graphite using everyday sticky tape, and won the Nobel Prize in physics in 2010 for their work. Scientists use creativity and imagination in their research as well as systematic and focused thinking. How does scientific research impact on our daily life?

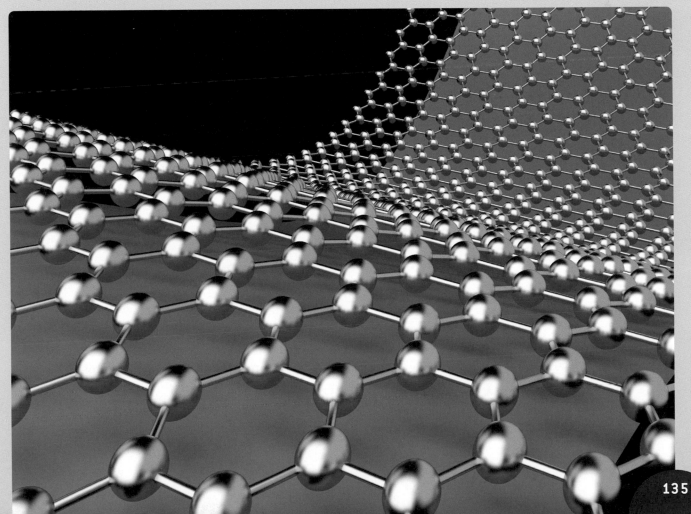

Key concept: Relationships

Related concept: Form

Global context: Identities and relationships

Introduction

The theories and laws of chemistry are fundamentally important in supporting the development and progress of fields as varied as physics, biology, astronomy, medicine, engineering, geology, microbiology, pharmacology, oceanography and climatology. All these areas of knowledge have important and inseparable relationships with the laws of chemistry. For this reason, chemistry is often referred to as the central science.

The concept of form includes the features of an object that can be observed, identified, described, classified and categorized. The nature of science informs us that experimental evidence can be collected using our senses. We make observations of the form, or features, of matter and the interactions matter undergoes during chemical reactions. Evidence collected by observation is used to develop theories about the nature of matter. Our ability to define and understand the form of matter helps us advance our understanding of the Universe.

Our relationship with the Universe shapes our development as individuals, impacting on our place in the world and what it is to be human. Technological developments over the last hundred years have affected our relationships both on a personal and a wider level. We need to be aware that technological advancement and the wellbeing of our communities and environment are deeply interconnected.

▼ Copper appears in many forms. Chemists need to have techniques they can use to identify them. How many different forms of sodium can you think of?

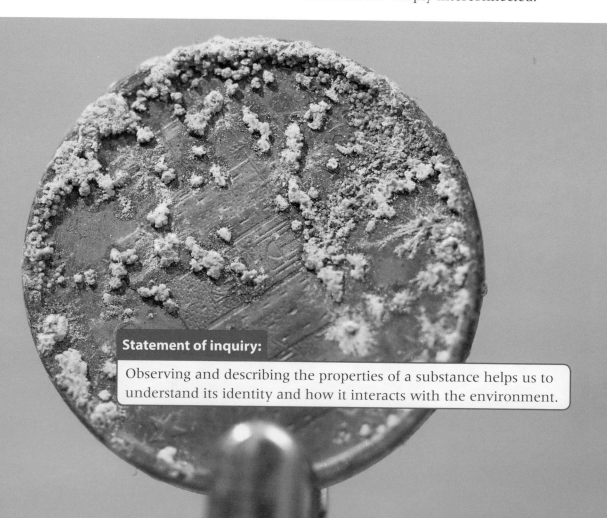

Statement of inquiry:

Observing and describing the properties of a substance helps us to understand its identity and how it interacts with the environment.

Due to the growth of scientific knowledge, we have a different awareness and understanding of our environment. While humans can be inaccurate in making observations, advances in technology have enabled instrumentation and sensors to perform far more precise observations than previously possible. However, the role of the scientist is still vital in the interpretation of the data collected, with the progression in our level of understanding being ultimately determined by human ability. For these reasons, the concepts of relationships, identities and form are explored in this chapter.

◀ The color of these transition metal complexes is determined by the identity of the metal, the charge of the metal ion and the nature of the species (or ligand) bonded to the metal ion to form the complex. This photo includes iron(III) chloride, nickel(II) nitrate, copper(II) sulfate, cobalt(II) chloride, and chromium(III) chloride. Can you match any of the salts with their names?

◀ The photos here, along with the photo on the previous page, show some of the common forms of copper. Can you identify what they are?

What are the properties of matter?

Matter is all around us and we are all made from matter. What is it and how can we identify it?

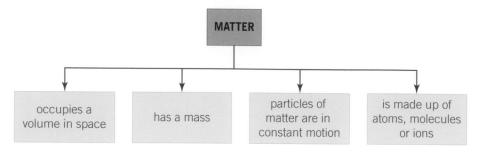

▲ What is matter?

Matter commonly exists in one of three states: solid, liquid and gas. This classification system helps us to understand and predict the behavior of matter.

		Solid	Liquid	Gas
Macroscopic properties	**Volume**	Fixed	Fixed	
	Shape	Fixed	Not fixed, takes the shape of its container	Not fixed, expands to occupy the space available
	Can it be compressed	No	No	Yes
Microscopic properties	**Forces between particles**	Attractive forces hold particles in close packed arrangement	Weaker than in solids	Tiny, treated as zero
	Motion of particles	Vibrate in fixed positions	Vibrate, rotate, and translate (move around)	Vibrate, rotate, and translate faster than in a liquid

What happens to the form of matter during a change of state?

In Chapter 4, Energy, we discussed the energy content of matter that exists above absolute zero (–273.15°C). We defined energy as the ability of an object to do work or transfer heat. Potential energy can be considered as stored energy. When the potential energy of matter is converted into kinetic energy, it undergoes movement. How does this relate to a change in state of matter?

1. Compare these values of enthalpies of fusion and vaporization. Explain why the enthalpy of vaporization for a substance is greater than its enthalpy of fusion.

	Enthalpy of fusion (J g^{-1})	Enthalpy of vaporization (J g^{-1})
Water	+334	+2,260
Carbon dioxide	+184	+574
Lead	+23	+871

2. Explain why the temperature of a boiling liquid does not continue to increase above its boiling point.

3. Steam burns are regarded as being far more harmful than a burn from boiling water. What change of state is occurring when steam reaches your skin? Why are the burns severe? Explain this in terms of energy transfer.

4. Explain why your skin feels cold when you leave the water after swimming on a hot summer day.

5. Why does a boiling liquid immediately stop boiling the moment you remove the source of energy?

Data-based question: Cooking at altitude

The boiling point of water is 100°C at an atmospheric pressure of 1 atmosphere or 100 kPa, a typical pressure at sea level. At higher altitudes the air pressure decreases and the boiling point decreases.

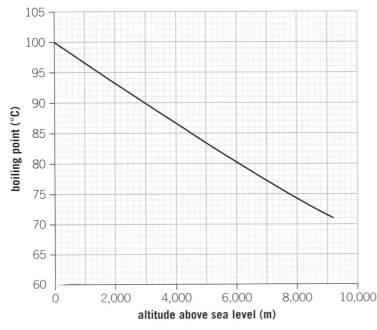

Landmark	Height (m)
Everest	8,848
Burj Khalifa	828

1. Study the graph and the table above. Use the information to calculate the percentage increase of the boiling point from the summit of Everest to the top of the Burj Khalifa.

2. Will it take less time to cook pasta at the top of the Burj Khalifa or the summit of Everest? Explain your answer. What do you expect to happen to the boiling point below sea level?

What is the relationship between vaporization and condensation?

In energy terms, evaporation is the opposite of condensation:

$$\Delta H_{vaporization} = -\Delta H_{consensation}$$

Energy is released from the system into the surroundings when water vapor begins to condense back into a liquid form. Similarly, melting, or fusion (going from solid to liquid), is the opposite in energy terms to the process of freezing (going from liquid to solid):

$$\Delta H_{fusion} = -\Delta H_{freezing}$$

Fusion and vaporization are endothermic processes—heat energy is transferred into the system from the surroundings. Condensation and freezing are exothermic processes—heat energy is released into the surroundings from the system.

1. Explain how the role of energy differs between condensation and evaporation.

▲ Steam is water in its gas phase. In fact, we only "see steam" from a boiling kettle because some of the gas particles condense to form small droplets of water

▲ This Eurasian crane's breath shows water vapor condensing when it is expelled into cold air. The condensed water droplets refract sunlight causing the tiny droplets to be visible

What is sublimation?

We have considered a substance changing state from solid to liquid to gas and back again, but under certain conditions some substances can change directly from solid to gas without passing through the intermediate liquid phase. This is called sublimation and is an endothermic process. The reverse process is called deposition and is exothermic:

$$\Delta H_{sublimation} = -\Delta H_{deposition}$$

1. At −78.5°C solid carbon dioxide (dry ice) sublimates. However, the white vapor you see is not carbon dioxide—it is a mist of water droplets condensed from the air. Explain the changes in state that occur.

▲ The misty effect of dry ice

What are the differences between a pure substance and a mixture?

Matter can be classified as solid, liquid or gas, but scientists have other ways to classify it as well. All matter can be classified as either a pure substance or a mixture. A mixture contains more than one substance; they are mixed together but not chemically combined.

▲ Sugar and iron shavings are both pure substances. In the centre is a mixture of the two

A pure substance contains only one substance. Matter that has a constant composition of atoms in an element, molecules in a covalent compound or formula units of an ionic compound are defined as pure substances. Pure substances melt and boil at fixed temperatures. When an impurity is present, melting and boiling occur over a range of temperatures.

How pure is pure?

In reality few of the substances that we describe as pure are 100% pure.

Ingredients: Purified Water, Sodium Chloride, Magnesium Sulphate, Potassium Chloride.	
TYPICAL ANALYSIS Servings Per Container 2.5 Calories 0	
PPM	
Total Dissolved Solids	<75
Magnesium	2
Sodium	17
Sulphate	7
Potassium	2
Chloride	32
Fluoride	<0.1
Calcium	1

	Typical value
Aluminium	<8.07 µg Al/l
Calcium	61.6 mg Ca/l
Residual chlorine - Total	0.59 mg/l
Residual chlorine - Free	0.54 mg/l
Copper	0.0726 mg Cu/l
Iron	10.7 µg Fe/l
Lead	<1.67 µg Pb/l
Magnesium	21.0 mg Mg/l
Manganese	<0.570 µg Mn/l
Nitrate	4.72 mg NO3/l
Sodium	26.1 mg Na/l

▲ Label from bottled water

▲ Analysis of typical British tap water

Scientists working with chemicals need to take care that contaminants do not affect the outcomes of their experiments. There are many grades of chemicals, but those of the highest level of purity are awarded an ACS rating. This standard is set by the American Chemical Society and chemicals must exceed the standards to be awarded this description. The term **reagent** is used to describe a chemical of a lesser purity that meets but does not exceed the ACS standard. There are lower grades for chemicals that are to be used in the production of foods and pharmaceuticals. School laboratory chemicals are classified as lab chemicals. These chemicals can be used effectively for your experiments but lack the purity required for the food, pharmaceutical and drug industries.

1. In your daily life, what other examples can you think of where the composition of a product is carefully analyzed and the percentage composition stated on the packaging?

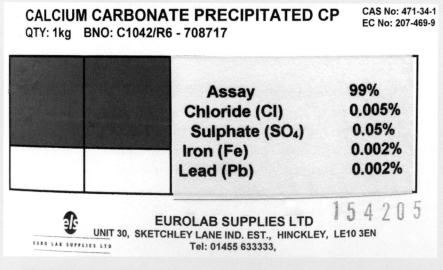

▲ The packaging labels on the chemicals that you use in your school laboratory provide you with a lot of useful information

Are all mixtures the same?

Mixtures consist of two or more chemical constituents that are combined physically rather than chemically. Salt water is a mixture, with the main components being the ionic compound sodium chloride and the molecular covalent compound water. The individual components of a mixture are not chemically bonded to each other, and can be separated by physical methods.

Imagine a bowl of fruit salad. If you did not like one of the fruits, you could just remove it. This is an example of physical separation. It is possible to separate the fruits because these constituents are not chemically bonded. Similarly, yellow sulfur powder mixed with iron filings can be separated out with a magnet as the iron is attracted to a bar magnet.

Mixtures like these where the composition varies are called heterogeneous. The composition of a heterogeneous mixture is not constant. Physical and chemical properties of this type of mixture will vary depending on the sample taken—two spoonfuls of fruit salad will contain a different proportion of individual constituents.

▲ Fruit salad

▲ Sulfur and iron shavings

Other mixtures are uniform; their composition is the same throughout. These are called homogeneous mixtures; many are referred to as solutions. Homogeneous mixtures are a physical combination of two or more chemical constituents which are evenly distributed throughout the mixture. As a result any sample taken from it will have the same physical and chemical properties. Vinegar is an example of a homogeneous mixture. It is composed of up to 20% acetic acid and water. Both liquids mix (or are miscible) with each other and the composition is consistent. Another example is salt dissolved in water. The parts of a mixture are not chemically bonded and can be separated by physical methods (see Chapter 7, Function).

▶ The features of a substance or mixture that can be observed and identified enable us to classify the type of matter

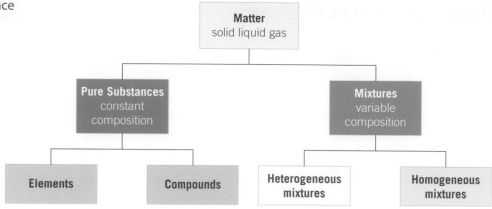

A phase is a system that is homogeneous in both physical and chemical terms. A phase is separated from another phase by a boundary layer.

When gases mix, regardless of how many components there are in the sample, they are said to be in one phase. For example, you cannot observe the presence of different individual gases in the atmosphere.

Liquids are different to gases.

● Two or more liquids that mix evenly with each other are said to be miscible; they exist in one phase. An example in the laboratory is the mixture of ethanol and water.

● When liquids do not mix with one another, they are referred to as immiscible liquids. They exist in separate phases and you can observe a phase boundary layer when they are left to settle. An example of this is oil and water.

▶ Oil and water are immiscible; they exist in separate phases

How does the size of particles change the type of mixture?

A solution is formed when a solid solute is dissolved in a liquid solvent. For example, if you place a teaspoon of salt granules into a glass of warm water, after mixing you will observe a colorless liquid with no evidence of the solid solute. You have created a homogeneous mixture.

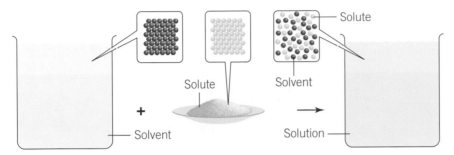

▲ Dissolving a solute in a solvent forms a homogeneous mixture

A solution is a homogeneous mixture. Any sample taken from it would have the same composition. It is impossible to see the particles under a microscope or even with an electron microscope, but what about mixtures with larger particles?

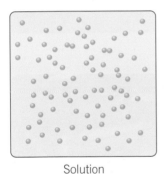

| Solution | Colloid | Suspension |

◄ The particles in a solution (a heterogeneous mixture) are much smaller than in a colloid or suspension (which are both homogeneous mixtures)

A suspension is a heterogeneous mixture in which the particles are much bigger than in a solution. In a suspension, we can often see the particles under a microscope. Some examples of a suspension are sulfur in water, and soil or silt in water. The particles tend to settle at the bottom of their container. They can be separated out using filtration or a centrifuge. Suspensions also exist in a gas; for example, dust in air.

◄ A suspension of soil settles when left to stand

A colloid is a mixture in which the particles are larger than those in a solution, but smaller than in a typical suspension. The particles are dispersed throughout the mixture and cannot be seen with a standard microscope. The particles will not settle on standing and they will pass through standard filter paper. They can be separated out using a centrifuge. Particles in a colloid can be seen if viewed with an electron microscope. Examples of colloids include aerosol sprays, milk and gelatin.

An emulsion is a type of colloid, a mixture of tiny particles that are suspended in a substance in which they will not mix or dissolve. The most common example of an emulsion is the mixture of oil and

▲ Fine spray from an aerosol

▲ When mixed, lead(II) nitrate, $Pb(NO_3)_2$, and potassium chromate, K_2CrO_4, produce soluble potassium nitrate, KNO_3, and lead(II) chromate, $PbCrO_4$, a yellow precipitate

▲ A centrifuge can be used to separate blood into red blood cells, white blood cells and plasma. In general, centrifuges can be used to separate out colloids and suspensions

▲ Flour and water form a colloid

An emulsion is a mixture of two or more insoluble liquids. Oil and water is one example of an emulsion. What are some other examples of emulsions you can find in the kitchen?

water. Oil and water are immiscible in each other; that is, they will not mix. When a mixture of oil and water is shaken together, the mixture that results is composed of tiny droplets of oil suspended in water. The mixture often has an opaque appearance.

Sometimes the colloidal particles suspended in a gas or liquid are seen to move about randomly—this is called Brownian motion. It is caused by the collision of colloidal particles with molecules of the fluid in which the particles are dispersed. It can be seen by observing particles of smoke through a microscope.

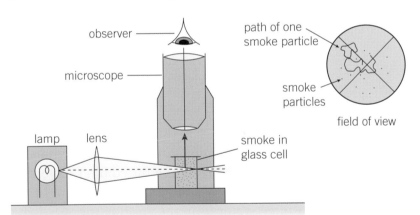

The particles of smoke are buffeted by the particles of air causing them to move around randomly and continuously

ATL Communication skills

Organize and depict information logically

Presenting information graphically can help to clarify your understanding of a topic and highlight interrelationships. For example, Lewis structure diagrams are used to enhance our understanding of how covalent bonding works in a compound (see Chapter 9, Models).

Develop a graphic to represent and clarify the meanings and connections between the following terms: solution, suspension, colloid, emulsion.

ANALYTICAL

How can we separate mixtures in the laboratory?

We can use our knowledge of the physical properties of the constituent parts of a mixture to design methods to separate them out. For example, filtration can be used to separate heterogeneous mixtures made up of insoluble solids and a liquid (usually water).

A double replacement reaction occurs between ions in solution. During the reaction, the cations or positive ions switch places and react with a different anion or negative ion. When one of the new ionic compounds or salts that form is insoluble in the solvent, a precipitate is observed.

Often, we want to separate the two products. How can we do this?

Experiment

Separating heterogeneous mixtures by filtration

Filtration can be used to separate the two new compounds. A fluted filter paper is sometimes used to increase the rate of filtration.

 Safety

- Wear safety glasses.
- Dispose of all residues appropriately.

Materials

- 0.5 mol dm^{-3} iron(III) nitrate
- 0.5 mol dm^{-3} sodium carbonate
- Standard filter paper
- Two test tubes and rack
- 25 cm^3 measuring cylinder
- 100 cm^3 beaker
- Dropping pipettes
- Filter funnel
- Distilled water in wash bottle

Method

1. Using a dropping pipette, transfer 10 cm^3 of 0.5 mol dm^{-3} iron(III) nitrate solution into a test tube and observe the color.

2. Using a dropping pipette, transfer 10 cm^3 of 0.5 mol dm^{-3} sodium carbonate solution into a test tube and observe the color.

3. Add these two solutions to the 100 cm^3 beaker at the same time. What do you observe? Allow the mixture to settle and observe the color of the solid and the solution separately.

4. Take a single filter paper and flute it ready for filtration.

5. Set up the filtration apparatus as shown in the diagram. Secure the fluted filter paper in the funnel by adding a small amount of distilled water from the wash bottle.

6. Swirl the solid–liquid mixture to mobilize the solid and slowly pour it onto the fluted filter paper. Pour slowly so the mixture doesn't flow over the top of the filter paper.

7. Wash the remaining residue from the beaker using distilled water from the wash bottle.

8. Dry the separated solid residue by placing the filter paper on a watch glass and leaving it in a warm place in the laboratory.

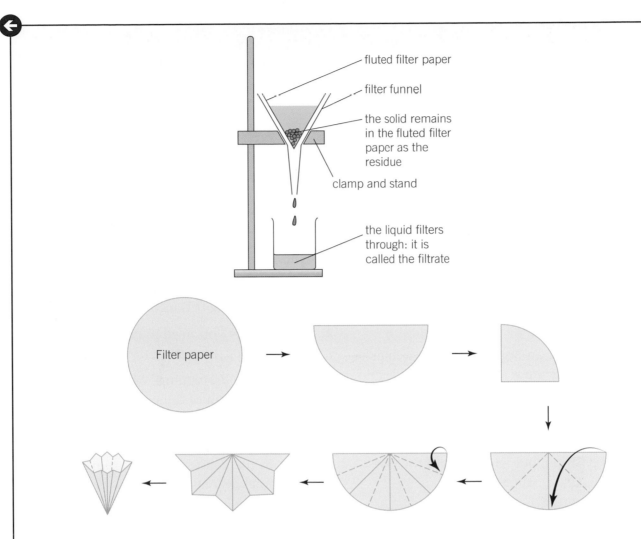

Questions

1. Write the word equation for the reaction between iron(III) nitrate and sodium carbonate.

2. Write a balanced chemical equation for the reaction.

3. Using your knowledge of chemistry, examine the products of this reaction and predict the composition of the insoluble salt. Explain your reasoning.

4. Which of the following mixtures could be separated using filtration?

    a) Sugar dissolved in water

    b) Sand and water

5. Explain how you could separate out the sand from a mixture of salt and sand.

As we have seen, liquids that do not mix with one another are referred to as immiscible. They exist in separate phases and you can observe the boundary layer when they are left to stand and are not being physically mixed, as we have seen with oil and water.

Experiment

Separating immiscible liquids

⚠ Safety

- Wear safety glasses.
- Cooking oil should not be disposed of down laboratory drains.
- Organic solvents should be disposed of appropriately.

Materials

- 250 cm³ separatory funnel
- Ring stand or retort stand with large clamp
- Distilled water
- Cooking oil or an organic solvent such as pentane or hexane

Method

1. Place the separatory funnel into the ring stand and check that the stopcock is in the closed position, perpendicular to the glass tubing.

2. Add approximately 75 cm³ of water and 75 cm³ of cooking oil.

3. Place the lid on the funnel and place the lid into the palm of your hand. With your other hand hold the stopcock in the closed position. Shake the flask until the liquids appear to be mixed.

4. Return the funnel to the ring stand and remove the stopper.

5. Over time make observations of the mixture as the liquids settle.

6. Place a 250 cm³ beaker below the separatory funnel and open the tap slowly to allow the lower layer to run into the beaker.

7. Close the tap when the lower layer has been separated.

Questions

1. What types of liquids are typically separated using this technique?

2. Explain why you need to remove the stopper prior to opening the stopcock.

3. From your observations, can you infer the composition of the upper layer of the mixture? Explain your reasoning.

4. Describe what happened to the boundary layer, as the volume of liquid remaining in the funnel decreased.

◀ A separatory funnel can be used to separate immiscible liquids. The narrowing cross-sectional area allows a relatively accurate point of separation to be achieved. This type of funnel is also used in the solvent extraction of substances in synthesis reactions

Crude oil, also known as petroleum, is formed over millions of years from decaying animal and plant material. It is a heterogeneous mixture of different types of hydrocarbons, sulfur, nitrogen, oxygen, metals and salts (see Chapter 2, Evidence). The composition of this mixture varies according to where it is extracted. To make use of crude oil the different types of hydrocarbons (fractions) need to be separated from each other.

The process of fractional distillation utilizes the differences in the boiling point of the various fractions of crude oil (see Chapter 2, Evidence). To separate the fractions the liquid crude oil mixture is heated until volatile fractions reach their boiling points and undergo a change of state from a liquid to a gas. The different fractions such as refinery gas, gasoline or petrol, naphtha, kerosene, diesel, fuel oil and bitumen can all be identified and categorized by their different boiling points. The industrial method for separating these fractions within crude oil is a simple and effective process.

▲ Crude oil is a viscous mixture of hydrocarbons

ⒶⒷⒸⒹ Experiment

Fractional distillation of crude oil

⚠ Safety

- Wear safety glasses.
- Crude oil substitute is highly flammable and harmful.

Materials

- Distillation apparatus (illustrated to the right)
- 2 cm³ of crude oil substitute
- Four small sample tubes
- Teat pipette
- 100 cm³ beaker
- Hard glass watch glass
- Wooden splint

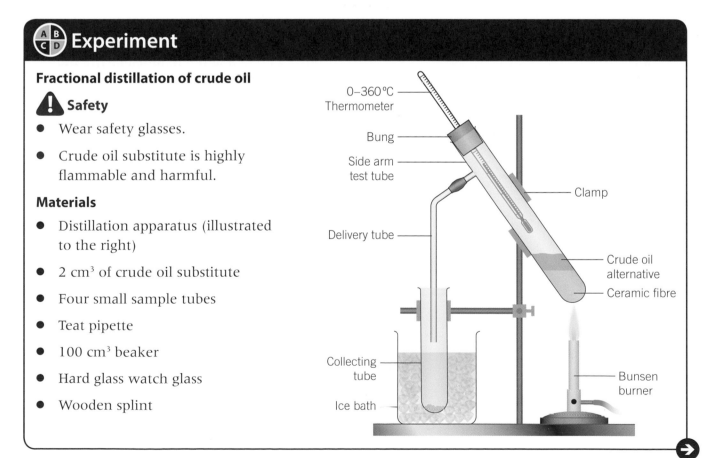

Method

1. Put about 2 cm³ of ceramic fibre in the bottom of the side-arm test tube. To this add about 2 cm³ of crude oil substitute using a teat pipette.

2. Set up the apparatus as shown in the diagram. The bulb of the thermometer should be level with or just below the side-arm.

   Note: before collecting the first fraction, place a beaker of ice water around the collecting tube.

3. Gently heat the bottom of the side-arm test tube with the lowest possible Bunsen flame.

4. Monitor the temperature. When the temperature reaches 100°C, replace the collection tube with an empty one. The beaker of cold water is no longer necessary.

5. Collect three further fractions, to give the following fractions:

   - room temperature to 100°C

   - 100–150°C

   - 150–200°C

   - 200–250°C.

   A black residue will remain in the side-arm test tube.

6. Test the four fractions for viscosity (how easily they pour), color, smell and flammability.

   - To test the smell, gently waft the smell towards you with your hand.

   - To test for flammability, pour onto a hard glass watch glass and light the fraction with a burning splint.

Questions

1. What do you observe when you compare the colors of each of the successive fractions? Is there a pattern?

2. Compare the flammability of each of the successive fractions. Which fraction is more flammable?

3. What is the relationship between the boiling point and the flammability of the fraction? Explain your reasoning.

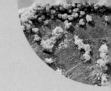

Summative assessment

Statement of inquiry:

Observing and describing the properties of a substance helps us to understand its identity and how it interacts with the environment.

Introduction

In this summative assessment, we will first look at how differences in the physical properties of substances can be utilized when selecting a method for the separation of homogeneous and heterogeneous mixtures. We then examine how analytical tests can help us to identify the composition of compounds. Finally, we will look at how the chemical composition of fatty acids can impact on the health of the community.

Separation of mixtures

Physical mixtures of elements and compounds can be separated by a variety of techniques including filtration, fractional distillation, evaporation, chromatography and using a separatory funnel. The substances in a physical mixture are not chemically bonded to each other.

Use the following information about three compounds (A, B, and C) to answer the questions.

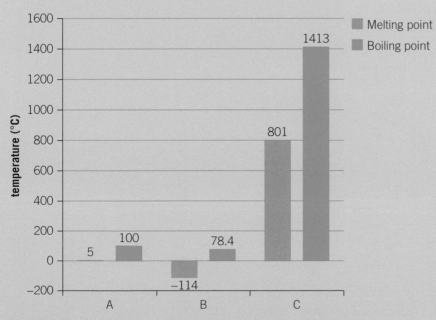

▲ Melting and boiling points of compounds A, B, and C

Substance	Solubility in water	Conductivity (solid)	Conductivity (liquid)
A	Yes	No	Yes
B	Yes	No	No
C	Yes	No	Yes

▲ Solubility, and conductivity of compounds A, B and C

1. Outline the following concepts. [2]

   a) Homogeneous mixture

   b) Heterogeneous mixture

2. State the most appropriate method for the physical separation of a mixture of compounds A, B and C. Justify your choice of technique. [6]

3. Determine the state of matter for each of the compounds A, B and C if the mixture is at a temperature of 80°C. Explain your reasoning. [3]

4. Predict the likely composition of substance C. Support your prediction with evidence. [3]

5. Identify the type of mixture containing compounds A, B and C. Explain your reasoning. [6]

The graph illustrates the changes in state for myristic acid, a saturated fatty acid, as it is heated. Saturated acids have been linked to high levels of cholesterol in humans and an increased chance of coronary heart disease.

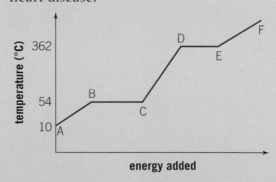

6. From your analysis of the graph:

   a) Explain what is happening between B and C. [3]

   b) State the name used to describe this process. [1]

   c) Describe what is happening to the particles in this pure substance between C and D. [2]

d) What is the boiling point of this liquid? [1]

e) The state of matter for this substance between E and F is a gas. Explain what will happen to the particles and the state of matter if energy is removed from the system. [3]

 Identify an unknown salt

As a result of high humidity in the chemical storeroom, the labels on five separate reagent bottles were lost. You will need to design a series of analytical tests to identify their composition.

| A | B | C | D | E |

A selection of analytical tests for cations and anions is summarized below.

Analytical tests for cations		
Cation	Addition of dilute sodium hydroxide	Addition of aqueous ammonia
Ammonia NH_4^+	Ammonia gas given off after warming	No reaction
Copper Cu^{2+}	Blue precipitate formed	Blue jelly-like precipitate formed dissolves in excess ammonia to form a deep blue solution
Iron Fe^{2+}	Green gelatinous precipitate formed	Green gelatinous precipitate formed
Iron Fe^{3+}	Rust brown gelatinous precipitate formed	Rust brown gelatinous precipitate formed
Lead Pb^{2+}	White precipitate formed that dissolves in excess NaOH(aq)	White precipitate formed
Zinc Zn^{2+}	White precipitate formed	White precipitate formed that dissolves in excess ammonia
Aluminium Al^{3+}	Colorless precipitate formed	Pale white precipitate formed

Tests for anions		
Anion	**Test**	**Observations**
Chloride ion Cl^-	A. Add 3 drops of dilute nitric acid followed by 3 drops of silver nitrate solution B. Add an excess of the ammonia solution to the resulting precipitate	A. A white precipitate is formed. B. The precipitate is soluble in ammonia
Bromide ion Br^-		A. A pale yellow precipitate is formed. B. The precipitate is slightly soluble in ammonia
Iodide ion I^-		A. A cream precipitate is formed. B. The precipitate is insoluble in ammonia
Sulfate ion SO_4^{2-}	Add 3 drops of barium chloride followed by 3 drops of dilute hydrochloric acid	White precipitate is formed
Carbonate ion CO_3^{2-}	Add hydrochloric acid dropwise	Effervescence—gas is produced

7. All of the salts are soluble in water. Design a series of tests to identify the composition of salts A–E. Your method should include a description of apparatus used and descriptions of the experimental techniques. [5]

8. Design a results' table to accompany your experimental design. [10]

Analysis and evaluation

A chemist's observations and results from these tests are shown below.

Testing for cations

(A) I added NaOH to unknown salt A. A white precipitate formed. On addition of ammonia, the white precipitate dissolved.

(B) Blue precipitate with NaOH added. Jelly substance dissolved with excess ammonia. Blue color deepened precipitate formed.

(C) Add NaOH.
• White solid formed.
• Add NH_3, solid dissolves slowly.

(D) NaOH ➔ Blue colored precipitate NH_3 added-jelly like substances seems to dissolve excess NH_3

Deepen of blue color.

(E) Add NaOH to solution. No reaction observed. Warmed it gently, strong smell of ammonia

Testing for anions

(A) No reaction with silver nitrate. White precipitate forms with the addition of barium chloride / H^+.

(B) Same observations as for unknown salt A... white ppt with barium chloride.

(C) White precipitate formed when acidified silver nitrate added. The precipitate completely dissolved when excess ammonia was added.

(D) 3 drops HNO_3 ➔ Precipitate forms. + $AgNO_3$ White/Yellow color

Add NH_3 - Little bit of the precipitate dissolves

(E) Beige colored precipitate after adding HNO_3 and silver nitrate

9. Present the data in an appropriate manner so that it can be used in a report. [3]

10. Identify the composition of the unknown salts A, B, C, D and E. [5]

The impact of palm oil

This article provides an overview of the benefits and problems of palm oil production.

Palm oil is one of the most widely consumed oils in the world. An ancient native tree of Africa, which is over 5,000 years old, it is now mostly grown commercially in south-east Asia. Palm oil is used for a range of purposes including frying and baking food stuffs and as biofuel, when blended with other fuels. The organic waste from processing the oil is also used as a biofuel, meaning that there is very little waste from the entire process. As the use of trans-fats in food manufacturing has become increasingly unpopular in the past few decades, due to the proven deleterious effects on people's health from increased cholesterol, the use of palm oil has sky-rocketed. However, palm oil itself may present health challenges due to the very high amount of saturated fat, specifically the 16-carbon saturated fatty acid, palmitic acid.

The vast palm oil plantations which now cover huge areas of Malaysia, Indonesia, and Thailand, to name a few countries, have led to the environmental problems of deforestation, smoke haze, and reduction in animal and plant species, as once varied landscapes have become farmed monocultures. Yet, the fact remains that palm oil can be produced economically and is therefore used globally in food manufacturing and as an alternative fuel source. It has also lifted hundreds of thousands of previously subsistence farmers out of poverty and contributes significantly to many countries' GDP (gross domestic product).

The article above provides an overview of the benefits and problems of palm oil production.

The iodine number of a fat is indicative of the level of saturation. The lower the iodine number, the greater the level of saturation and the more harmful the oil is for humans. Other sources of fats for food manufacturing include olive oil (iodine number 79–95), lard (iodine number 43), and soya bean oil (iodine number 125–145). Palm oil has an average iodine number of 48–58. Saturated fatty acids lead to an increase in the level of cholesterol in the blood stream, and an increased risk of coronary heart disease.

11. Summarize how our increased scientific understanding of the impact of human development on the environment enables us to identify the issues that exist with palm oil production. [3]

The features, and physical and chemical properties of different types of fats and oils are better understood as a consequence of the application of scientific research. Our ability to identify and analyze the chemical structure of substances we consume helps scientists to identify the connections between form and function.

12. Examine the iodine number of different oils used in food manufacturing and household cooking and determine which is considered the most healthy to consume. Explain the reasons for your answer. [3]

7 Function

▲ The Grand Prismatic Spring in Yellowstone National Park is the largest hot spring in the United States. The bright colors around the perimeter are due to photosynthetic pigments within bacteria. The centre of the spring at 93°C is too hot for the bacteria clustering on the cooler perimeter, but billions of organisms called thermophiles live in the scalding water. Organisms adapt to their surroundings so that they can function and survive. What other plants and animals have adapted to their environment?

▲ Fireflies are in fact a luminous beetle. They function by producing light but no heat from an internal organ. The chemical reaction between oxygen and the chemical referred to as luciferin produces the pulsing light. It used to be thought that the purpose of the light was to deter predators, but it is now thought that it is used in selecting a mate. What other living organisms function by producing light?

▲ This robot has been designed to deal with explosive devices. What features enables it to carry out this function?

◀ This panther chameleon is one of over 150 different species of chameleon. It has a spectacular way of responding to its environment. Changes in skin color can be in response to danger, anger, fear, temperature or humidity. The male can use changes in color to show dominance to other males and to attract a female for mating. Rather than a chemical reaction, the chameleon changes the structure of special skin cells called iridophore cells. This in turn leads to a change in the color of their skin. What other animals use changes of color?

Key concept: Relationships

Related concept: Function

Global context: Globalization and sustainability

Introduction

The ability to separate mixtures is an important skill for chemists. When several elements and compounds are present in a mixture, scientists need to understand the physical and chemical properties of the substances involved and how these substances function with each other. For this reason the key concept of this chapter is relationships.

Mixtures are often separated using physical methods which rely on the different properties of the components in the mixture. For example, water purification, a vital process for society, uses a physical separation technique known as filtration. This method takes into account the relative sizes of impurities in order to remove them.

In this chapter we will learn about the mole. A mole of any substance always contains the same number of particles (atoms, molecules, ions, electrons): a mole of water contains 6.02×10^{23} molecules and a mole of copper contains 6.02×10^{23} atoms of copper. This number is known as Avogadro's constant and its function is to relate the number of particles found in a chemical substance to various units that we use to measure quantity, such as molar mass, concentration, mass and volume.

In order to maximize the use of available resources, we require an understanding of how the quantities of reactants influence the outcome of a reaction. This is a matter of concern both in the school laboratory and in an industrial production plant, as chemical processes use finite resources and can have an impact on the environment. For this reason, the global context of this chapter is globalization and sustainability.

▼ Winnowing is an ancient method of separating grain from its protective casing (known as chaff). When throwing the mixture into the air, the chaff, which is lighter than the grains, is blown away by the wind, while the grains fall back and are collected. This separation technique relies on the difference in mass of the two components

Statement of inquiry:

The way in which matter functions is dependent on its properties and the relationship of the different systems within the environment.

People have long wanted to know the composition of things. Assaying is the analysis of metal ores to find the proportions of their components. This illustration shows an assayer at work. It appeared in *The Laws of Art and Nature, in Knowing, Judging, Assaying, Fining, Refining and Inlarging the Bodies of Confin'd Metals* which was published in 1683. Why might you want to know the composition of a metal?

▶ This broken glass is mixed with spilt sugar. Understanding the different properties of the glass and sugar allows them to be separated. How would you separate them?

◀ The balloon contains 1 mole of air. The beaker contains 1 mole of nickel(II) chloride, and the flask contains 1 mole of copper(II) sulfate in 1 litre of water

How can soluble and insoluble salts be separated?

The physical separation techniques used in a laboratory are selected depending on how the components of the mixture function. The techniques that can be used to separate the constituents depend on whether the mixture is homogeneous or heterogeneous, and the physical properties of the constituents (see Chapter 6, Form).

Imagine a jar filled with jelly beans. This is an example of a heterogeneous mixture. If you don't like the licorice ones, you can simply remove them from the jar. This is an example of physical separation. When the mixture contains constituents that are in the same phase, such as a mixture of different liquids, understanding the relationships and differences in the physical properties of the constituents helps to identify the best process of physical separation.

Ionic compounds (see Chapter 5, Conditions) are generally formed from the electrostatic attraction between a cation, normally a metal, and an anion, normally a non-metal. The combination of most cations and anions result in the formation of a salt. The following solubility rules enable you to predict whether a salt is soluble in water. Each cation and anion functions in a different way in aqueous solution.

▲ No two handfuls of these jelly beans will be the same. This is an example of a heterogeneous mixture

Soluble	Insoluble
All nitrates	None
Most sulfates	Lead sulfate and barium sulfate
Most chlorides, bromides and iodides	Silver chloride, silver bromide, silver iodide, lead chloride, lead bromide and lead iodide
Sodium carbonate and potassium carbonate	Most other carbonates
Sodium hydroxide and potassium hydroxide	Most other hydroxides

▲ Simplified solubility rules

A double-replacement reaction occurs when the cations from two different ionic compounds replace each other within a compound. A simple general equation helps to explain this:

$$A^+ B^- + C^+ D^- \rightarrow A^+ D^- + C^+ B^-$$

In this example, the cations (the positively charged ions) are A and C, and the anions (the negatively charged ions) are B and D.

- If AD and CB are both salts that are soluble in aqueous solution, no solid precipitate is formed.

- If one of the salts formed is insoluble in aqueous solution, then a precipitate is formed and the physical separation technique of filtration can be used to separate the solid from the aqueous solvent.

Consider the reaction between solutions of sodium chloride and silver nitrate:

sodium chloride + silver nitrate → sodium nitrate + silver chloride

$$NaCl(aq) + AgNO_3(aq) \rightarrow NaNO_3(aq) + AgCl(s)$$

The solubility rules confirm that the white precipitate formed is silver chloride.

Filtration is a simple but effective separation technique that utilizes the differences in how substances function in solvents.

▲ Silver chloride precipitate has a characteristic milky-white appearance

Experiment

Testing for the presence of halide ions

 Safety

- Dilute nitric acid is corrosive.
- Wear safety glasses.
- Avoid contact with the skin.
- Silver solutions should not be flushed down the sink; dispose of them as chemical waste.

Materials

- $0.4\,mol\,dm^{-3}$ nitric acid solution
- $0.5\,mol\,dm^{-3}$ silver nitrate solution
- $0.5\,mol\,dm^{-3}$ sodium chloride solution
- $0.5\,mol\,dm^{-3}$ sodium bromide solution
- $0.5\,mol\,dm^{-3}$ sodium iodide solution
- Three test tubes and a test tube rack
- $10\,cm^3$ measuring cylinder
- $100\,cm^3$ beaker
- Dropping pipette

Method

1. Place $5\,cm^3$ of the sodium chloride, sodium bromide and sodium iodide solution into three separate test tubes.

2. In a separate beaker, prepare the silver nitrate solution by acidifying it with dilute nitric acid.

3. Using a dropping pipette, transfer 2–$3\,cm^3$ of the acidified silver nitrate into each of the three test tubes.

4. Record your observations, paying particular attention to any color changes.

Questions

1. Predict the products of each chemical reaction by constructing a word equation.

2. Using the solubility rules, state the salt that forms the insoluble precipitate.

3. Analyze the colors of the precipitates formed and describe the trend in color of the halide salt as you move down Group 17 from chlorine to bromine to iodine.

1. Predict the state of matter of each of the products of the following reactions:

    a) lead(II) nitrate + sodium sulfate → lead(II) sulfate + sodium nitrate

    b) calcium bromide + sodium carbonate → calcium carbonate + sodium bromide

    c) barium chloride + silver nitrate → barium nitrate + silver chloride

    d) potassium bromide + silver nitrate → potassium nitrate + silver bromide

    e) sodium iodide + silver nitrate → sodium nitrate + silver iodide

Extension: Construct the balanced chemical equations for each of the reactions above.

The components of a solution are:

-   the solvent—the liquid into which the solid will dissolve

-   the solute—the solid substance to be dissolved

-   the solution—the resulting combination of solvent and solute, combined in a single phase.

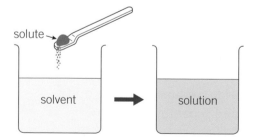

▶ Components of a solution

What makes some salts insoluble in water? Some molecules are polar (see Chapter 9, Models), and some are not. A general rule for the formation of solutions is **like dissolves like.**

This means that:

- polar solvents will dissolve polar solutes
- non-polar solvents will dissolve non-polar solutes
- polar solvents will not dissolve non-polar solutes.

In covalent bonding, atoms share electrons. When a covalent molecule is made up of identical atoms, the bonding electrons are shared equally between the two atoms. An example of this is an oxygen molecule. These molecules are described as being non-polar.

In a water molecule, oxygen shares electrons with each of the hydrogen atoms. The oxygen has a much greater attraction for these shared electrons than the hydrogen atoms, resulting in an uneven distribution of the bonding electrons in the covalent bond. Overall this molecule is described as being polar and will dissolve other polar compounds.

Polar water molecules are attracted to the charged ions present in a crystal of sodium chloride. The partial positive charge on the hydrogen atoms in a water molecule attract the chloride anions while the partial negative charge on the oxygen atom attracts the sodium cations. The result is that sodium chloride dissolves in water.

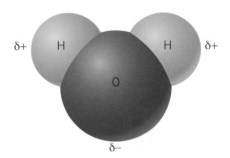

▲ An oxygen molecule is non-polar as the bonding electrons are shared equally between the two atoms

▲ Water molecules are polar

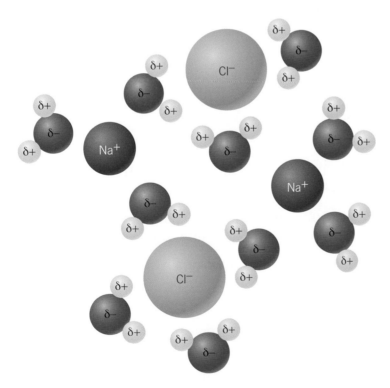

◀ Sodium chloride dissolves in water because of the attraction between the water molecules and the sodium and chloride ions

Filtration relies upon differences in the solubility of the substances in the mixture. If you mix sand and water, the sand is insoluble and can easily be separated from the water by filtration. However, if you have a mixture of sodium chloride and water, the sodium chloride dissolves in the water and filtering the mixture will not result in separation.

Experiment

The reaction between lead(II) nitrate and potassium iodide solutions

Ionic compounds vary in their solubility in water, which is a polar solvent. How can we separate compounds that have differences in their solubility?

 Safety

- Wear safety glasses.

- Lead nitrate is toxic and hazardous to the environment. Dispose of this chemical and the residues of the filtration appropriately.

Materials

- $0.01\,mol\,dm^{-3}$ lead(II) nitrate solution

- $0.5\,mol\,dm^{-3}$ potassium iodide

- Two $25\,cm^3$ measuring cylinders

- $50\,cm^3$ beaker

- Filter paper

- Distilled water

- Apparatus shown below

- Watch glass

- Disposable gloves

Method

1. Measure $10\,cm^3$ of $0.01\,mol\,dm^{-3}$ lead(II) nitrate solution using a $25\,cm^3$ measuring cylinder.

2. Measure $10\,cm^3$ of $0.5\,mol\,dm^{-3}$ potassium iodide solution using a $25\,cm^3$ measuring cylinder.

3. Place a $50\,cm^3$ beaker on top of a white surface, such as a piece of white paper or a white tile.

4. Combine the two colorless solutions by pouring the contents of the two measuring cylinders into the beaker at the same time.

5. Observe what happens on the initial mixing of the two liquids and then again after 10 minutes.

6. Prepare a fluted filter paper (see Chapter 6, Form) and set up the apparatus shown below.

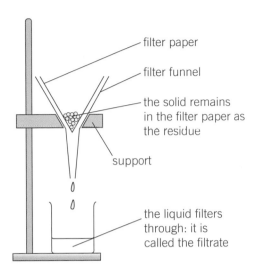

7. Swirl the beaker to mobilize the precipitate and slowly pour the mixture into the filter paper so the mixture does not flow over the top of the filter paper. Rinse the flask with distilled water to ensure all the precipitate is filtered.

8. Wearing gloves, remove the filter paper and lay it flat on a watch glass. Leave in a warm place to dry slowly.

Questions

The balanced chemical equation for this reaction is:

$$Pb(NO_3)_2(aq) + 2KI(aq) \rightarrow 2KNO_3(aq) + PbI_2(s)$$

1. What observations did you make? Were they what you expected?

2. What term do we use to describe this type of reaction?

3. After 10 minutes, what did you observe? What was the color of the precipitate that formed?

4. In this type of reaction we often refer to the presence of spectator ions. These are ions that do not play a direct part in the reaction. Focusing on the formation of the precipitate lead(II) iodide, what are the spectator ions in this reaction?

5. Following filtration, what is the color of the lead salt you have made?

6. Why are the characteristic colors of transition metal compounds useful for identifying compounds?

How can miscible liquids be separated?

Liquids that mix together and form a homogeneous mixture are said to be miscible. An example of this is ethanol and water, which are both polar molecules. When liquids mix and form a homogeneous mixture, the method of physical separation used must take into account how they function and their physical properties. When you observe a mixture of ethanol and water, can you see any distinct layers of separation of the two liquids?

A homogeneous mixture of miscible liquids with different boiling points can be separated by using distillation. Water has a boiling point of 100°C and ethanol 78.37°C. In simple distillation, the mixture is slowly heated. Ethanol reaches its boiling point first and begins to undergo a change in phase from liquid to gas (evaporation). This gas travels through the water-cooled condenser where it condenses back to a liquid. The liquid can be collected and isolated. What remains is a pure solution with the higher boiling point, in this case water.

▶ Simple distillation apparatus used to separate two miscible liquids with different boiling points. Remember to use a heating mantle or hot plate if you are heating ethanol. Never use a Bunsen burner as ethanol vapor is highly flammable

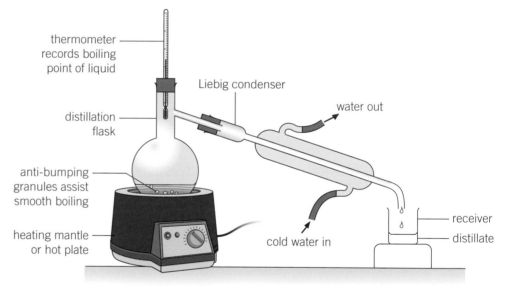

thermometer records boiling point of liquid

Liebig condenser

water out

distillation flask

anti-bumping granules assist smooth boiling

heating mantle or hot plate

cold water in

receiver

distillate

ᴬᵀᴸ Information literacy skills

Finding, interpreting, judging and creating information

You may have already done a simplified form of fractional distillation (see Chapter 6, Form). Research standard laboratory fractional distillation and answer the following questions.

1. Compare the similarities and differences between simple distillation and fractional distillation.

2. Explain what happens to the boiling points of the individual fractions when they are combined in a mixture.

3. When performing simple distillation of a three-part mixture, the distillate fractions are not considered to be pure after the first distillation. Explain why this is the case.

Why are some liquids immiscible?

Oil and water are said to be immiscible as they do not mix. As oil is less dense than water it floats on the top of it, forming a heterogeneous mixture in which the two distinct layers are evident—remember the rule "like dissolves like". Oil molecules are non-polar and water molecules are polar.

▲ If oil is released accidentally at sea it floats on the top of the water and can be washed up on the shoreline. This can have a devastating effect on the natural environment

▲ The instructions on a bottle of salad dressing tell you to shake it well before use to temporarily distribute the components to form an emulsion

A separatory funnel (see Chapter 6, Form) is an effective method of separating immiscible liquids. With a narrowing cross-sectional area towards the tap, a relatively accurate point of separation can be achieved. This type of funnel is also used in the process of solvent extraction.

How can chromatography separate mixtures?

Chromatography is an important analytical chemical technique which is widely used in forensic science. There are many forms of this analytical technique, but paper chromatography and thin-layer chromatography are most commonly used in the school laboratory. They are used in the separation of the components of a heterogeneous mixture.

Chromatography is versatile; the type of chromatography used can be chosen to suit the functions and properties of the substances analyzed. For instance, gel permeation chromatography is often used to analyze very large molecules, or polymers.

The principle behind all forms of chromatography is that there is a stationary phase (a solid or a liquid held on a solid) and a mobile phase (a liquid or a gas). The mobile phase flows through the stationary phase and carries the substance under test with it. Different components travel at different rates.

In paper chromatography, the stationary phase is the chromatography paper and the mobile phase is a solvent. The choice of solvent is important as the components of the substance under test must be soluble in it.

 Experiment

Are there common pigments in different types of food coloring?

Materials

- Chromatography paper
- 250 cm³ beaker
- Distilled water
- Four different food colorings
- Ruler and HB pencil

Method

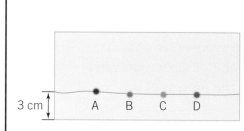

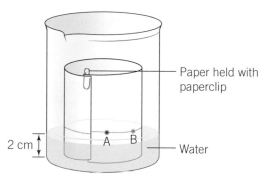

1. Cut a piece of chromatography paper into a rectangular shape, so that when it is rolled up it fits inside the beaker without touching the sides.

2. Draw a pencil line with a ruler 3 cm from the bottom of the paper.

3. Place distilled water in the beaker to a depth of 2 cm, no more.

4. Lay the chromatography paper flat on the laboratory bench and add a single drop of each of the food colorings along the pencil line. Label each dot so that you know which coloring is which. Allow time for the liquid to absorb into the paper before placing it in the beaker.

5. Make observations as the solvent front rises up the chromatography paper. Just before the solvent front reaches the top of the paper, remove it from the beaker, lay it out flat and allow to dry.

6. With a pencil, mark the position of the different pigments and measure the distance travelled by each component.

7. Measure the maximum distance the solvent travelled.

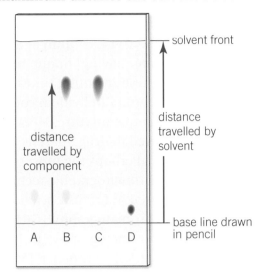

You can calculate the retention factor for each pigment by using the following rule:

$$\text{retention factor } R_f = \frac{\text{distance travelled by a component}}{\text{distance travelled by the solvent}}$$

Calculate the retention factor of each distinct pigment for each food coloring.

Questions

1. Examine your results. Are there any pigments that are common to two or more food colorings? How did you decide that they were the same pigment?

2. What does the retention value tell you about the solubility of the separated component? Explain your reasoning.

3. What effect would a different solvent have on the results of this experiment? Compare your ideas with others.

ATL Information and media literacy skills

Locate, evaluate, synthesize information from a variety of sources

The following article, taken from The Forensics Library website, details how different forms of chromatography can be applied to forensic science. Read the text carefully and then answer the questions.

As a very basic analytical test, planar chromatography can be applied to the analysis of inks and dyes. An analyst may need to establish whether two pieces of writing were written using the same ink, and chromatography is one way of, at the very least, acting as a presumptive test for this. The ink samples will be spotted onto the separation medium and allowed to travel through the matrix. Two different ink samples may travel at a different rate through the stationary phase, thus indicating that they may not be from the same source. The same principle can be applied to dyes. For example, banknotes may be marked with dye which, if this gets onto the hands of a criminal (eg a thief), chromatography may be used to indicate if the dye on the hands of a suspect is the dye used to mark the stolen money under scrutiny. Of course further analysis would be necessary to confirm this.

High-performance liquid chromatography is often utilized in the analysis of materials used to make explosives. If dealing with a substance that is suspected to have been used in the production of an explosive device, HPLC can be used to provide qualitative analysis to aid in the identification of the material. Different substances will travel down the HPLC column at a different rate, having unique retention times. These retention times can be used to identify individual components in a sample mixture. HPLC can also be used in drug analysis.

Perhaps the most confirmative chromatographic technique used in forensic science is gas chromatography [GC] or liquid chromatography [LC] coupled with mass spectrometry [MS], which can allow for substances to be identified. Libraries of mass spectra may be used to aid the analyst in identifying a sample, along with their own expertise. The technique will often be used alongside other tests. For instance, in the analysis of suspected illicit substances, presumptive tests may be used to give the analyst a clue as to the kind of drug they are dealing with. But these tests do not confirm identity, thus LC-MS and GC-MS may then be used to provide a more certain identification.

1. How could chromatography be used to investigate the legitimacy of two wills left by a deceased person?

2. How could the police use HPLC to investigate a series of terrorist bombings to see if they are related?

3. International level sport is a very lucrative business for participants, managers, and sponsors. How can the World Anti-Doping Agency (WADA) use forensic techniques such as chromatography in their attempt to keep the use of performance-enhancing drugs in sport to a minimum?

4. Research recent illegal doping at the Olympic level. Compile a list of countries and sports which have been identified by WADA as being connected to "drug cheats".

The first table gives paper chromatography results for five known ions. The second table shows the R_f values and spot colors for three samples of waste water.

Ion	Molar mass (g mol⁻¹)	Distance travelled (cm)	R_f	Color of spot
Nickel, Ni²⁺	58.7	0.8	0.08	Pink
Cobalt, Co²⁺	58.9	3.5	0.35	Brown-black
Copper, Cu²⁺	63.5	6.0	0.60	Blue
Cadmium, Cd²⁺	112.4	7.8	0.78	Yellow
Mercury, Hg²⁺	200.6	9.5	0.95	Brown-black

Sample	R_f value	Color of spot
1	0.60	Blue
	0.78	Yellow
2	0.35	Brown-black
	0.95	Brown-black
3	0.08	Pink
	0.78	Yellow
	0.95	Brown-black

1. Identify the ions present in Sample 2.

2. Identify the ions present in Sample 3.

3. Analyze the data and list the ions in the first table from the fastest to the slowest.

4. Sketch the appearance of the chromatography paper of sample 1 after the analysis has been completed.

5. What do you observe about the relationship between the molar mass and the retention factor?

Why are moles important to chemists?

QUANTITATIVE

In his book *The Assayer*, Galileo Galilei said, "Mathematics is the language of science". Scientists analyze data collected during experiments to find patterns and mathematical relationships between variables.

The law of definite proportions states that a chemical compound always contains a fixed ratio of elements by mass. This law had its origins in the work of the French scientist Joseph Proust whose experiments proved the relationships between reacting elements.

▶ Joseph Proust stated in 1794 that chemical compounds combine in a fixed ratio. For example, the chemical compound water is made up of the same fixed proportions of hydrogen and oxygen

A mole of substance is defined by the IUPAC as 6.02×10^{23} particles of that substance. These particles could be atoms, molecules, ions or electrons. This value of 6.02×10^{23} is called Avogadro's constant. The substance might be an element, an ionic or covalent compound, or electrons produced during electrolysis (see Chapter 10, Movement). The symbol used to denote moles is n and the unit symbol is mol. Avogadro's constant allows us to examine the mathematical relationship between reacting species in a balanced chemical reaction.

ATL Critical thinking skills

Consider ideas from multiple perspectives

The magnitude of the Avogadro constant is difficult to relate to. To gain perspective, we can build up our understanding of size gradually.

- How many people are in your family?

- How many friends do you have at school?

- How many students are there in your school?

- How many people live in the city where you live?

- How many people live in your country?

- How many million kilometres is the Sun from the Earth?

- How many people are there on the planet?

Clearly the numbers are getting larger!

Imagine a small beaker that contains $18\,\text{cm}^3$ of water. How many water molecules do you think are in this beaker? It contains 1 mol of water molecules or 602,000,000,000,000,000,000,000 molecules of water!

▲ The great wildebeest migration that happens annually in the Serengeti of Tanzania and the Masai Mara of Kenya involves over 1 million animals. It is hard to imagine, yet this number is small when compared to Avogadro's number

One mole of some common elements. The graduated cylinders contain (left to right) 200.59 g mercury, 207.2 g lead, and 63.546 g copper. The flask on the left contains 32.006 g sulfur, and the other flask contains 24.305 g magnesium. The watch glass holds 51.996 g chromium. Everything rests on sheets of aluminum foil with a total mass of 26.98 g

Amedeo Avogadro (1776–1856) proposed that equal volumes of any gas measured at the same temperature and pressure contain the same number of molecules; this is known as Avogadro's law.

Imagine several balloons. At standard temperature and pressure (STP), each balloon is filled with a mole of a different gas. The relationship that exists between these balloons is interesting. Each balloon:

- contains Avogadro's number of atoms or molecules of its gas

- occupies an identical volume of 22.7 dm³

- has a different mass that is equivalent to the molar mass of its gas.

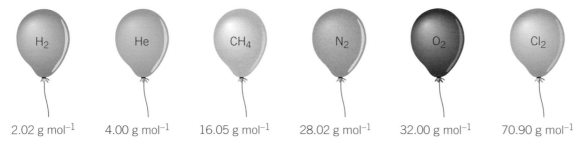

H_2	He	CH_4	N_2	O_2	Cl_2
2.02 g mol⁻¹	4.00 g mol⁻¹	16.05 g mol⁻¹	28.02 g mol⁻¹	32.00 g mol⁻¹	70.90 g mol⁻¹

▲ The molar volume of any gas is identical at a given temperature and pressure

What happens if the temperature of all the balloons is increased? The number of molecules and the mass of gas in each balloon stays the same, but the volume of each balloon increases by the same amount.

Standards

Standard temperature and pressure (STP) is defined as:

- standard temperature = 273 K or 0°C

- standard pressure = 100 kPa or 1 atmosphere

- molar volume of a gas at STP = 22.7 dm³ mol⁻¹.

Another common standard is room temperature and pressure (RTP) which is defined as:

- room temperature = 298 K or 25°C

- room pressure = 100 kPa or 1 atmosphere

- molar volume of a gas at RTP = 24 dm⁻³

1. Explain why the RTP volume is greater than the STP volume.

Avogadro's law simplifies our understanding of how gases function and the mathematical relationships between reacting gases in industrial processes. This is exploited in the Haber process, which is used to produce ammonia gas (see Chapter 1, Balance).

QUANTITATIVE Are all atoms of the same element identical?

Isotopes are atoms of the same element that have the same number of protons in the nucleus but a different number of neutrons. Many elements have naturally occurring isotopes; for example, carbon has three, chlorine has two and calcium has six. Isotopes of an element have very similar chemical properties, which means that they function in the same manner in chemical reactions. For example, hydrogen has three isotopes that occur naturally.

Isotope	Hydrogen-1	Hydrogen -2	Hydrogen-3
Alternative name	Protium	Deuterium	Tritium
Number of protons	1	1	1
Number of neutrons	0	1	2
Symbol	^1_1H	^2_1H	^3_1H
Structure ⊕ proton ◯ neutron			
Relative atomic mass A_r	1	2	3
Relative abundance	99.9885 %	0.0115 %	Negligible

▲ Comparison of the three isotopes of hydrogen

The mass number of an element that you see on the periodic table is the average weight of the atomic masses of the isotopes, weighted with respect to their relative abundances. This is known as the *relative atomic mass* of an atom, A_r.

Naturally occurring copper consists of 69.2% copper-63 (^{63}Cu) and 30.8% copper-65 (^{65}Cu). We can calculate the relative atomic mass A_r by working out the average atomic masses:

$$A_r = (63 \times 0.692) + (65 \times 0.308) = 63.616$$

Chlorine has two isotopes, chlorine-35 and chlorine-37, with relative abundance of 75% and 25% respectively:

$$A_r = (35 \times 0.75) + (37 \times 0.25) = 35.5$$

Some periodic tables use rounded numbers for the relative atomic mass while others state a more accurate number.

The *relative molecular mass* (or *relative formula mass*), M_r, is calculated by combining the A_r for the individual atoms or ions in a compound. To do this you need to know the formula so that you know the number of atoms or ions. The unit is g mol^{-1}.

Follow these steps to calculate the relative molecular mass of a compound.

1. Determine the composition of the compound, then work out the number of atoms of each element.

2. Find the value A_r for each element present in the compound from the periodic table. These are included in the periodic table.

3. To calculate the relative formula mass, multiply each A_r by the number of atoms present and calculate the sum of these values.

Molar mass

The molar mass of a substance is equal to the relative molecular mass or atomic mass. The unit for molar mass is g mol^{-1}.

Worked example: Calculating molar mass

Question

Calculate the molar mass of 1 mol of sulfuric acid H_2SO_4.

Answer

Element	Formula	Number of atoms	A_r	M_r
Hydrogen	H	2	1	$2 \times 1 = 2$
Sulfur	S	1	32	$2 \times 32 = 32$
Oxygen	O	4	16	$2 \times 16 = 64$
				Total 98

So the molar mass is 98 g mol^{-1}. In this example the relative atomic mass numbers have been rounded.

---

1. Calculate the molar mass of the following compounds using the values of A_r found in the periodic table.

a) CO_2

b) SO_3

c) HNO_3

d) Na_2CO_3

e) $KMnO_4$

f) $CaCl_2$

g) $Al(NO_3)_3$

h) $Fe_2(SO_4)_3$

i) $(NH_4)_2SO_4$

j) $Na_2S_2O_3.7H_2O$

QUANTITATIVE # What is the function of the mole?

Stoichiometry is the branch of chemistry that examines the relative quantities of reactants and products in a chemical reaction. These amounts are described in a number of ways:

- mass of solids in grams (g)

- volume of a liquid (cm^3) and its concentration ($mol\,dm^{-3}$)

- volume of a gas (dm^3)

- the number of representative particles.

▶ Relationship between number of particles, number of moles and mass

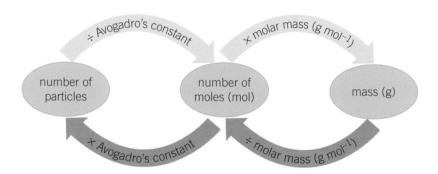

Converting between the number of particles, moles and the mass of a substance involves simple mathematical calculations which focus on proportions. For example:

- what proportion of a mole is a certain mass of a compound?

- what proportion of a mole is a certain number of particles?

Every chemical reaction can be described by a balanced chemical equation. This describes the stoichiometric relationship between reactants and products (see Chapter 1, Balance). The coefficients used to balance a chemical equation represent the mole ratio for the reaction. For example, consider the complete combustion of methane:

$$CH_4(g) + 2O_2(g) \rightarrow CO_2(g) + 2H_2O(g)$$

This chemical equation describes a reaction where 1 mol of methane CH_4 reacts with 2 mol of oxygen to produce 1 mol of carbon dioxide and 2 mol of water vapor.

To calculate the number of moles for a specific mass of a substance, the following equation is used:

$$\text{number of moles n (mol)} = \frac{\text{mass } m \text{ (g)}}{\text{molar mass } M_r \text{ (g mol}^{-1})}$$

Worked example: Moles and mass

Question

Calculate the number of moles in 4.00 g of sodium hydroxide (NaOH). Use the values of A_r from the periodic table.

Answer

$$\text{Number of moles } n = \frac{\text{mass } m \text{ (g)}}{\text{molar mass } M_r \text{ (g mol}^{-1})}$$

$$= \frac{4.00}{(22.99 + 16.00 + 1.01)}$$

$$= 0.100 \text{ mol}$$

When you consider the number of atoms, molecules, ions or formula units in a given sample, the relationship between Avogadro's constant and the number of representative particles is calculated as follows:

$$\text{number of moles } n = \frac{\text{number of particles}}{\text{Avogadro's constant}}$$

Worked example: Moles and particles

Question

Calculate the number of moles of methane (CH_4) in a sample of 1.20×10^{24} molecules.

Answer

$$\text{Number of moles } n = \frac{\text{number of particles}}{\text{Avogadro's constant}}$$

$$= \frac{1.20 \times 10^{24}}{6.02 \times 10^{23}}$$

$$= 2.00 \text{ mol}$$

Avogadro's law states that equal volumes of any gas measured at the same temperature and pressure contain the same number of molecules. At STP (273 K and 100 kPa), the molar volume of a gas is 22.7 dm³ mol⁻¹.

Worked example: Moles and volume

Questions

1. Calculate the number of moles found in a 13.62 dm³ sample of methane gas at STP.

2. Calculate the mass in grams of this sample of methane gas.

Answers

1. 1 mol of methane occupies 22.7 dm³ at STP

   $$\text{number of moles } n = \frac{13.62}{22.7}$$

   $$= 0.600 \text{ mol}$$

2. $$\text{Number of moles } n = \frac{\text{mass } m}{\text{molar mass}}$$

   Rearranging to make mass as the subject of the equation:

   $$m = n \times \text{molar mass (CH}_4)$$

   $$= 0.600 \times [12.01 + (4 \times 1.01)]$$

   $$= 9.63 \text{ g}$$

In the following questions use the values of A_r given in the periodic table.

1. Calculate the number of moles in each of the following masses:

   a) 9.8 g of sulfuric acid, H_2SO_4

   b) 25.0 g of calcium carbonate, $CaCO_3$

   c) 8.0 g of sodium hydroxide, NaOH

   d) 60 g of glucose, $C_6H_{12}O_6$

2. Calculate the mass (in grams) in each of the following:

   a) 0.25 mol of carbon dioxide, CO_2

   b) 3 mol of ammonia, NH_3

   c) 0.710 mol of calcium phosphate, $Ca_3(PO_4)_2$

   d) 0.211 mol of iron(III) oxide, Fe_2O_3

3. Calculate the number of particles present in the following:

   a) 9.80 g of phosphoric acid, H_3PO_4

   b) 60.77 g of iron(II) sulfate, $FeSO_4$

   c) 12.10 g of ethanoic acid, $C_2H_4O_2$

   d) 6.40 g of sulfur dioxide, SO_2

4. Calculate the volume of each of the following samples of gases at STP:

   a) 7.6 g of argon, Ar

   b) 100 g of ethene, C_2H_4

5. Magnesium burns in the presence of oxygen to form the metal oxide, magnesium oxide. This oxidation reaction is commonly performed in a school laboratory. A student weights out 18.0 g of magnesium for the reaction. The equation for the reaction is as follows:

   $$Mg(s) + O_2(g) \rightarrow MgO(s)$$

   a) Balance the chemical equation.

   b) Calculate the number of moles of the reactant, magnesium.

   c) If oxygen is in excess and all the magnesium is used up, calculate the number of moles of magnesium oxide formed.

   d) What mass of magnesium oxide is produced?

6. What mass of barium sulfate would be produced when 10 g of barium chloride is completely consumed in the following reaction?

   $$BaCl_2(aq) + H_2SO_4(aq) \rightarrow BaSO_4(aq) + 2HCl(aq)$$

7. Copper sulfate reacts with sodium hydroxide to produce a precipitate of copper hydroxide.

   $$CuSO_4(aq) + 2NaOH(aq) \rightarrow Cu(OH)_2(s) + Na_2SO_4(aq)$$

   a) Calculate the mass of sodium hydroxide needed to convert 15.95 g of copper sulfate into copper hydroxide.

   b) Calculate the mass of copper hydroxide produced.

Summative assessment

The way in which matter functions is dependent on its properties and the relationship of the different systems within the environment.

Introduction

In this summative assessment, you will first apply your conceptual understanding of separation techniques to experimental data. Secondly, you will design several scientific investigations to separate different types of mixtures. You will then reflect on the validity of experimental methods used in a number of distinct case studies. Finally, you will learn more about how desalination is becoming a practical application of science to address the global issue of increasing demand for water.

Separating the components of sea water

Student A is given the task of designing an experiment to separate and collect the components of a sample of sea water, NaCl(aq). They are provided with standard laboratory apparatus. Following the experiment, the student recorded the following data for analysis.

Initial volume of sea water = 200.00 cm³
Volume of water recovered = 196 cm³
Mass of salt NaCl recovered = 68.8 g

The student in their research for a laboratory report discovered the following facts.

	Sodium chloride	Water
Molar mass (g mol⁻¹)	58.44	18.02
Solubility in water (g/100 cm³)	35.9	-
Boiling point (°C)	1413	100
Appearance	White crystalline substance	Colorless liquid

1. Discuss which separation technique the student would have used to separate the sodium chloride from water. Give reasons for your choice of technique. [3]

2. Use the experimental data to determine the amount (in grams) of sodium chloride that was dissolved in 100 cm³ of the salt water sample. [3]

3. A saturated solution contains the maximum amount of a solute dissolved for a given solvent at a specific temperature. For this experiment, this amount is 35.9 g/100 cm³. State if the solution of sea water used by the student is a saturated solution. Show all working to support your answer. [3]

4. Calculate the number of moles of sodium chloride that was recovered by this separation process. [2]

5. Suggest reasons why all of the water was not collected during this separation process. [3]

6. Student B, in the same class, only managed to collect a sample of the solid sodium chloride. With the technique chosen by this student, the water is lost to the surroundings.

a) Describe the technique used by student B. [3]

b) Offer ideas why it is sometimes important to recover the solvent from a solution. [2]

 ## Separating mixtures

The American Chemical Society defines analytical chemistry as "the science of obtaining, processing, and communicating information about the composition and structure of matter. In other words, it is the art and science of determining what matter is and how much of it exists".

Simple separation techniques can be used in the school laboratory to separate substances from one another and help identify the composition of a mixture. Specific analytical tests for the presence of cations and anions can be used to identify the composition of ionic salts.

This investigation has been broken up into three parts.

1. **Mixture A** is a combination of three different miscible alcohols.

2. **Mixture B** is a mixture of four different solids.

3. Three different ionic salts, are labelled **A, B** and **C**. You are told the possible composition of the salts. You have to identify the contents of A, B and C.

Mixture A

Name	Molecular formula	Boiling point (°C)
Methanol	CH_3OH	65
Ethanol	C_2H_5OH	75
Propanol	C_3H_7OH	97

Mixture B

Name	Chemical formula
Calcium carbonate powder	$CaCO_3$
Sodium chloride	NaCl
Pebbles	Sandstone / Granite / Shale

Unknown compounds A, B and C

Three white ionic solids are labelled Solid A, B and C. The identity of each compound is unknown. They could be one of the following compounds:

● Barium iodide

● Potassium bromide

● Sodium chloride

Additional information

Element	Formula of ion	Flame color
Lithium	Li^+	Red
Sodium	Na^+	Orange
Potassium	K^+	Lilac
Barium	Ba^{2+}	Pale green
Copper	Cu^{2+}	Blue green

Flame tests are used in the identification of cations.

Halide ion	Color observed
Cl⁻	White precipitate
Br⁻	Pale cream precipitate
I⁻	Pale yellow precipitate

The addition of silver nitrate solution to halide ion solution results in the formation of insoluble salts, each with a characteristic color.

7. Design a series of experiments to:

a) Separate the individual fractions of the alcohol mixture and identify the composition of the fraction. [5]

b) Separate and recover the three solids and confirm the identity of sodium chloride and calcium carbonate by performing a number of simple tests. [5]

c) Identify solids A, B and C. [5]

Evaluating the choice of methods

8. The behavior of substances under different conditions, how they function, can be investigated to determine the most appropriate method of physical separation technique. Examine the following investigations and using your knowledge of the physical and chemical properties of the substances in the mixture, evaluate the validity of the method chosen to separate the components of the mixture. In each case:

● State the validity of the technique chosen. Is it the most appropriate technique?

● Explain improvements or extensions to the method of separation. [6]

Investigation	Mixture	Method of separation	Student scientific reasoning
A	Fabric dye	Paper chromatography	Individual components of liquid mixtures can be separated using the process of chromatography
B	Activated carbon and ethanol	Simple distillation / evaporation	Activated carbon is insoluble in water and most other solvents. Simple distillation will result in the removal of water, leaving the carbon as a residue
C	Benzene (non-polar) and water (polar), liquids	Filtration	As these two different liquids are not miscible, the denser liquid will be trapped by the filter paper while the other liquid will pass through and be separated.

Solar power and tomato farming

This article appeared on the *MIT Technology Review* website dated October 2016. Read the passage and then answer the questions.

A Desert Full of Tomatoes, Thanks to Solar Power and Seawater

Sundrop Farms has used a clever system to grow food using unlikely ingredients—but is the idea likely to catch on?

by Michael Reilly and Jamie Condliffe, October 6, 2016

At first glance, growing fruit in the desert sounds like an awfully good way to feed a mushrooming global population and adapt to the worst effects of climate change. And a farm in South Australia run by the greenhouse developer Sundrop Farms is doing just that, using solar power to desalinate water and grow tomatoes in the otherwise parched landscape. *Farmers Weekly* reports that the $150 million facility focuses sunlight from 23,000 mirrors onto a tower to produce energy that drives an attached desalination system. Sucking water from the nearby Spencer Gulf, it produces up to one million liters of fresh water every day.

The result is tomatoes—lots of tomatoes. The farm fills eight trucks every day with greenhouse-raised tomatoes, and it's expected to produce more than 15,000 tons of the things per year when it reaches full capacity.

But while it's certainly an impressive feat, there is an argument that says there's simply not much point using solar energy production facilities like this to grow fruit and vegetables. Talking to *New Scientist*, Paul Kristiansen from the University of New England in Australia said that it was "a bit like crushing a garlic clove with a sledgehammer." He added, "We don't have problems growing tomatoes in Australia."

He has a point. Desalination is on the rise in many areas of the world, but it is mostly done through reverse osmosis, an expensive and energy-intensive process. As a result, desalination plants really only make sense in places that are water-stressed and have the resources necessary to build, run, and maintain them.

That may change as cheaper techniques become more robust and fresh water supplies dwindle in some places as a result of climate change. Even then, though, this isn't a technology that's likely to catch on in poorer regions—which, unfortunately, are also the most likely to have a hard time adapting to shifting climate patterns. Solving that problem is likely to be a matter not of sheltering plants in greenhouses, but of designing crops that really can survive in a desert.

9. Identify and list the science and technology being used by Sundrop Farms in South Australia to grow tomatoes. [2]

10. Determine the problem that this technology is being used to solve. [1]

11. Despite the reported great success of this project, there is criticism of the final outcome—growing tomatoes. Evaluate the proposition made in *New Scientist* by Paul Kristiansen from the University of New England and debate with your peers if this criticism is justified. [4]

12. The writers of this article suggest that selective breeding and genetic modification of crops will be more practical in poorer, arid regions. What are their arguments to support this? [2]

13. Explain why this project is an example of how science and technology can alter the natural relationships that exist between food crops and the natural environment. [2]

14. What is your opinion of the Sundrop Farms project in South Australia? Evaluate the benefits and potential downsides of such a project. Structure your answer as a summary that would enable comparison with your peers. [4]

8 Interaction

The gravitational interactions between the Earth, Moon, and Sun take place in an ordered system that controls the tides. Although the Earth's oceans are constantly changing, our understanding of these gravitational interactions on the Earth's oceans means that tides are a forecastable feature of our lives. However, the consequences of these interactions are sometimes destructive. Why is it important for communities to be able to predict the patterns of the tides?

The interaction between remora or sucker fish and sharks is known as symbiosis. The remora fish travel with the shark, cleaning off parasites and eating unwanted remains from the shark's diet. Their interaction is mutually beneficial. Do you know any other symbiotic relationships in the animal and plant kingdoms?

Rapid globalization has resulted in some countries developing very quickly. Infrastructure such as transportation systems can become stressed and unable to adequately service the needs of the community. Indonesia is an emerging market economy that has undergone rapid economic growth. It is now experiencing overwhelming traffic congestion. Does society need to control the pace of change to make these interactions sustainable?

▲ Global population growth and changes in fishing techniques have resulted in overfishing, to the point where some species are no longer self-sustaining. Since the early 1990s there has been a moratorium on fishing for cod in the North Atlantic Ocean following the collapse of fish stocks. Recovery has been slow due to the depleted food supplies, a lack of large breeding females and changes in the aquatic ecosystem due to climate change. What other examples are there of our interaction with a natural resource threatening its sustainability?

REDOX

What is a redox reaction?

There are three main types of reactions studied by chemists.

- acid–base reactions

- precipitation reactions

- reduction-oxidation (redox) reactions.

Many reactions that you have already studied are redox reactions, but you have perhaps not yet defined them in this way.

A redox reaction is the interaction between two different processes: reduction and oxidation. These process are interdependent—one cannot occur without the other.

- Oxidation is a reaction in which a substance loses electrons, gains oxygen or loses hydrogen.

- Reduction is a reaction in which a substance gains electrons, loses oxygen or gains hydrogen.

▶ Redox reactions are all around us. An enzyme called polyphenol oxidase reacts with oxygen in the air to oxidize the apple. Iron, the main component in steel, oxidizes in atmospheric air if left unprotected, decreasing the structural integrity of objects made from steel. The combustion of methane is an oxidation reaction resulting in the release of large amounts of heat and carbon dioxide. Iron ores, such as hematite, contain iron(III) oxide; the iron can be extracted by reducing it in a blast furnace in a series of reactions. This results in the removal of the oxygen from iron(III) oxide

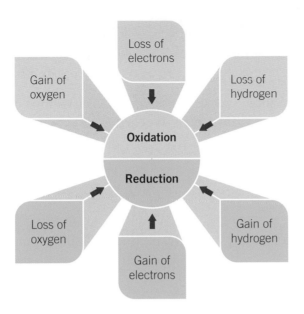

Loss of electrons

Gain of oxygen

Loss of hydrogen

Oxidation

Reduction

Loss of oxygen

Gain of hydrogen

Gain of electrons

What is oxidation?

REDOX

Oxidation reactions are perhaps the most familiar examples of redox reactions as we experience them all around us. A common example is the rusting or corrosion of iron and steel objects. The element iron (the main component in steel) is oxidized to produce iron(III) oxide, more commonly referred to as rust:

$$\text{iron} + \text{oxygen} \rightarrow \text{iron(III) oxide}$$
$$4\text{Fe}(s) + 3\text{O}_2(g) \rightarrow 2\text{Fe}_2\text{O}_3(s)$$

This chemical process is irreversible and the global annual cost is estimated to be US\$2.5 trillion. This is over 3% of the world's gross domestic product (GDP). Public and private organizations spend large amounts of time and money developing and implementing methods of preventing and managing corrosion.

1. Identify examples of corrosion in your school, home or community. Describe the impact of corrosion. How can it be prevented or reduced?

◄ The Brooklyn Bridge in New York City is one of the oldest bridges in America. Its construction began in 1870. Structures made from steel require constant preventative maintenance to limit the damage caused by the process of corrosion or rusting. How would your life be affected if infrastructure in your closest city was irreparably damaged by corrosion?

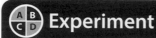 **Experiment**

The corrosion of metals

Different metals corrode at different rates when exposed to oxygen. Observations of the corrosion process made over time help us to classify metals as reactive or unreactive.

 Safety

- Wear safety glasses.

Materials

- Small samples of magnesium, aluminium, zinc, iron, nickel, tin, lead and copper
- Nine test tubes and a test tube rack
- 250 cm³ beaker
- Distilled water
- Bunsen burner, tripod stand and heatproof mat, or kettle
- Parafilm

Method

1. Place one or two small pieces of each of the metals into eight separate test tubes. Label each test tube with the symbol of the metal in it and put it in the test tube rack. Half-fill each test tube with distilled water.

2. In the ninth test tube place a piece of iron and label appropriately. Place the test tube into the test tube rack. Bring a small quantity of water to the boil. Fill this test tube to the top with the boiling water; then seal it with Parafilm. Your teacher can show you how to do this.

3. Make observations of each test tube then set the test tube rack aside for two weeks.

4. After two weeks, observe the metal in each test tube for evidence of corrosion.

Questions

1. What did you notice about the rate of corrosion for the metals tested?

2. Place them in order from most corroded to least corroded.

3. What did you observe in the test tube with the iron and boiled water?

4. Explain any differences in the reactions between the iron in the boiled water compared to that in the normal distilled water. What does this tell you about the required conditions for corrosion of a metal?

Data-based question: How widespread are the effects of metal corrosion?

Metal corrosion has been a problem since the Iron Age. Research into corrosion prevention has made significant progress, but the problems it causes are still very significant.

This data is the result of a two-year study on the impact of corrosion in some sectors of the US economy at the turn of the century.

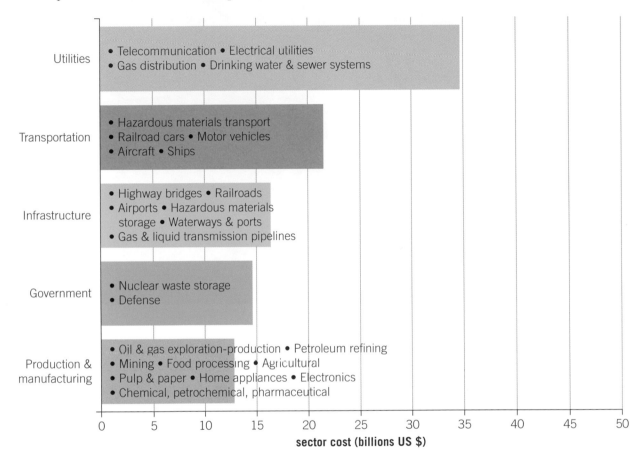

1. Which sector of the economy is impacted the most by corrosion?

2. In a group, brainstorm why this sector is most affected by corrosion, then summarize the main reasons.

3. The next highest cost sector is transportation. Explain the possible reasons for this.

4. Manufacturers are investing in technology to reduce the amount of corrosion in motor vehicles. What strategies are they using?

5. Ships are protected from corrosion by a sacrificial anode. Research this term. What are the most common metals used for these anodes?

▲ Sacrificial anodes on the rudder and propeller of a fishing trawler

6. Describe some other simple corrosion prevention techniques used by industry.

REDOX

Can oxidation reactions be useful in terms of energy?

Alkanes are a type of hydrocarbon; they contain only carbon and hydrogen atoms. They have the general formula C_nH_{2n+2} (see Chapter 2, Evidence). When they combust (burn in the presence of oxygen) they release large amounts of energy.

The combustion of alkanes in oxygen is a redox reaction. It is an exothermic reaction, so in an open system, heat energy is released into the surroundings. In the presence of sufficient atmospheric oxygen, the alkanes undergo a process called complete combustion. The global economy and our day-to-day existence are highly dependent on these exothermic combustion reactions.

In many countries, liquid petroleum gas (LPG) has become a common fuel for motor vehicles such as taxis, public buses, freight vehicles and, increasingly, private passenger vehicles. It is produced by refining crude oil. It is commonly a mixture of the hydrocarbon gases, propane, C_3H_8, and butane, C_4H_{10}.

The combustion of butane gas is described by the equation:

$$2C_4H_{10}(g) + 13O_2(g) \rightarrow 8CO_2(g) + 10H_2O(g) + \text{heat}$$

The combustion of 1 mole of butane releases 2878 kJ. So reacting 5 g of butane produces enough energy to bring 740 g of water at 20°C to the boil at 100°C. Check this calculation is correct by using the equation $E = mc\Delta T$. You will remember this equation from Chapter 4, Energy.

Alternatives to gasoline and diesel

The combustion of alkanes such as butane in LPG and heptane, C_7H_{16}, in gasoline (see Chapter 2, Evidence) releases large amounts of carbon dioxide, contributing to the global increase in greenhouse gases and global warming. This interaction with the natural environment is one of the most significant ways we make an impact on our planet. As the number of traditional internal combustion vehicles on the roads increases so does our impact on the global environment. However, increasing levels of carbon dioxide are not the only concern.

Gasoline and diesel are a complex mixture of alkanes, alkenes, and other components. When they combust, carbon monoxide, nitrogen oxides and particulates are produced as well as heat and carbon dioxide. They are poisonous in their own right and contribute to the formation of smog.

◀ This heavy smog in Moscow is typical of the pollution that results from the large number of vehicles in many cities around the world

1. Since 2006, all public taxis in Hong Kong run on LPG. How do LPG-powered vehicles help to reduce the level of pollution?

Information literacy skills

Process data and report results

The following report is from *Business Insider*. It considers predictions that the global number of cars on the road and the number of kilometres flown in planes will nearly double by 2040. It is based on research by a company called Bernstein.

The number of cars worldwide is set to double by 2040

Cars are projected to reach the two billion mark by 2040, while air travel kilometers are set to hit 20 trillion in the same period.

Bernstein said it expects most of this transport growth to happen in emerging markets like China and India, as global populations are set to rise by another two billion over the next 25 years to 9.2 billion.

Growing GDP in those regions will increase demand for items once seen as luxuries, including automobiles and flights.

The fastest growth will be in air travel, according to the report, which is particularly sensitive to per capita GDP growth. As *Business Insider* previously reported, this will support increased use of oil in emerging markets over the next two decades, countering the drop in demand from developed economies.

Here's a visual of projected global transport growth over the next 25 years:

Number of cars billion	2015	1.1 billion cars
	2025	1.5 billion cars
	2040	2.0 billion cars
Number of trucks million	2015	377 million trucks
	2025	507 million trucks
	2040	790 million trucks
Air revenue Passenger km trillion RPK	2015	9 trillion RPK
	2025	12 trillion RPK
	2040	20 trillion RPK

While this growth looks large, the report notes that demand in the US and Europe will fall as both regions reached peak oil [consumption] over ten years ago in 2005.

"Given the peak in travel speeds, high penetration of transport, and increasingly services-led nature of economic growth in the West, demand from this area will decline from here," it added.

"Although China (gasoline demand) in particular will remain the most important market over the next decade, India (diesel demand) is set to become the most important market over the next 25 years."

Germany is planning to ban the sale of petrol-fueled vehicles from 2030. France and the UK will do the same and include diesel-fueled vehicles from 2040. According to the World Health Organization (WHO), France, Germany and the United Kingdom presently have approximately 43 million, 52 million and 36 million registered vehicles respectively. China has over 250 million and India 160 million.

The arguments for and against limiting access to fossil fuel sources, technologies and development in many forms have been ongoing for many decades. Developing economies may not have the resources—financial, technical, or natural—to adopt cleaner technologies.

1. "Developing economies lead the way in reducing carbon emissions and fossil fuel usage. Their role in the planet's future is crucial."In teams of four–six students, debate this statement using the information above to help. Half of you should defend the statement and the other half should oppose it.

How does the supply of oxygen affect the combustion of alkanes?

The combustion of alkanes requires oxygen. The amount of oxygen available affects the type of combustion reaction and consequently the products. When an excess of oxygen is available (as in the combustion of butane that we looked at earlier), this is called complete combustion, and the products are carbon dioxide and water.

Bunsen burners often burn methane (although some use butane, propane or a mix of both gases). Complete combustion takes place when you open the gas sleeve and a blue, roaring flame is produced:

$$CH_4(g) + 2O_2(g) \rightarrow CO_2(g) + 2H_2O(g) + heat$$

If the sleeve is closed the amount of oxygen available is reduced and incomplete combustion occurs. This results in the production of carbon monoxide instead of carbon dioxide:

$$2CH_4(g) + 3O_2(g) \rightarrow 2CO(g) + 4H_2O(g) + heat$$

Carbon monoxide is very poisonous.

If even less oxygen is available, the carbon does not combine with oxygen and a black solid—soot—is produced:

$$CH_4(g) + O_2(g) \rightarrow C(s) + 2H_2O(g) + heat$$

	Ratio of methane to oxygen
Complete combustion producing carbon dioxide and water	1:2
Incomplete combustion producing carbon monoxide and water	2:3 or 1:1.5
Incomplete combustion producing carbon (soot) and water	1:1

◀ Comparing the complete and incomplete combustion of methane

▲ The blue flame indicates that complete combustion is occurring. Sufficient oxygen is being supplied and the flame burns cleanly. The yellow flame indicates that incomplete combustion is occurring. Insufficient oxygen is being supplied and the flame does not burn cleanly. What do you see on the bottom of any glassware heated over a yellow flame?

The dangers of carbon monoxide

Carbon monoxide is a colorless, odourless and tasteless gas that is very poisonous to humans. When it enters the bloodstream, it binds to the hemoglobin molecules in the red blood cells, blocking the site where oxygen and carbon dioxide normally bind. This lowers the oxygen carrying capacity of the blood. Symptoms of carbon monoxide poisoning include feeling dizzy, followed by nausea, vomiting and tiredness, eventually leading to a coma and death.

Carbon monoxide poisoning can be caused by poorly maintained combustion engines that produce higher levels of carbon monoxide. It is treated by the inhalation of pure oxygen at higher than normal atmospheric pressure; this displaces the carbon monoxide molecules attached to the hemoglobin molecules.

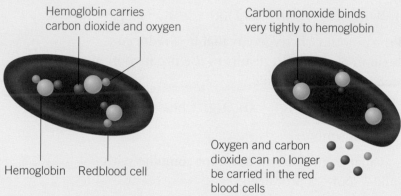

Hemoglobin carries carbon dioxide and oxygen

Carbon monoxide binds very tightly to hemoglobin

Hemoglobin   Redblood cell

Oxygen and carbon dioxide can no longer be carried in the red blood cells

1. How might you prevent the formation of carbon monoxide during the combustion of alkanes?

Experiment

Complete and incomplete combustion

Alcohols are compounds with a wide variety of structures and applications. They are commonly used in industry for synthetic reactions. During this experiment, you will perform some simple combustion reactions and evaluate your observations to classify the different reactions as either complete or incomplete combustion reactions.

 Safety

- Wear safety glasses.
- Alcohols are flammable and harmful liquids. Store them carefully away from the combustion reaction.

Materials

- Methanol, ethanol, propan-2-ol and butanol
- Four porcelain crucibles
- Large beaker filled with tap water or a damp cloth for extinguishing the flame
- Lighter
- Four test tubes

Method

1. Pour 10 cm³ of one of the alcohols into a crucible.

2. Remove all bottles and beakers of alcohol from the bench.

3. Light the alcohol with the lighter and observe the color of the flame.

4. When the flame is burning strongly, place the test tube into the main section of the flame for up to 20 s, securing it in this position with a clamp.

5. Look for any evidence of discoloration on the test tube.

6. Repeat this procedure for the other alcohols.

7. Record your observations in a table.

Questions

1. Draw the structural formula for each alcohol in your table.

2. Analyze the qualitative data collected and then explain with reasons why each reaction is either complete or incomplete combustion.

3. What is the composition of the substance that appeared on the test tube?

4. How does increasing the carbon-length of alcohols affect the type of combustion reaction that occurs?

5. Write balanced chemical equations for each combustion reaction. Base the products for each reaction on the conclusions you reached about the nature of the combustion reaction.

What are the products when a metal reacts with an acid?

When a metal reacts with an acid, a salt and hydrogen gas are produced (see Chapter 3, Consequences):

$$\text{metal} + \text{acid} \rightarrow \text{salt} + \text{hydrogen}$$

The reaction between magnesium and hydrochloric acid is a common reaction which you may have already encountered. The magnesium displaces the hydrogen ion from solution and hydrogen gas is released. The hydrogen can be collected and tested with a lit splint, producing a popping sound.

$$Mg(s) + 2HCl(aq) \rightarrow MgCl_2(aq) + H_2(g)$$

We can predict the composition of the salt from the metal and the acid. The metal (Mg in this example) provides the positively charged cation (Mg^{2+}) of the salt and the acid (HCl) provides the negatively charged anion (Cl^-) or suffix of the salt. This reaction is classified as a single replacement reaction.

Acid	Metal	Salt	General rule
hydrochloric acid HCl	magnesium Mg	magnesium chloride $MgCl_2$	hydrochloric acid produces a chloride salt
sulfuric acid H_2SO_4	zinc Zn	zinc sulfate $ZnSO_4$	sulfuric acid produces a sulfate salt
nitric acid HNO_3	iron Fe	iron(III) nitrate $Fe(NO_3)_3$	nitric acid produces a nitrate salt
ethanoic acid CH_3COOH	aluminium Al	aluminium acetate $Al(CH_3CO_2)_3$	ethanoic acid produces an acetate salt

▲ The composition of the product from the reaction between a metal and an acid can be predicted

REDOX

Can the reactivity of metals be ordered?

The reactivity series is an experimentally derived list that orders metals from the most to the least reactive (see Chapter 3, Consequences). It can be used to predict the interactions in a single replacement reaction.

Metals above hydrogen in the reactivity series can displace the hydrogen ion from solution in the reaction between a metal and an acid. The rate of the reaction is dependent on the position of the metal relative to hydrogen.

▶ The reactivity series of metals can be used to predict the products of redox reactions. Why is gold and silver used to make jewelry?

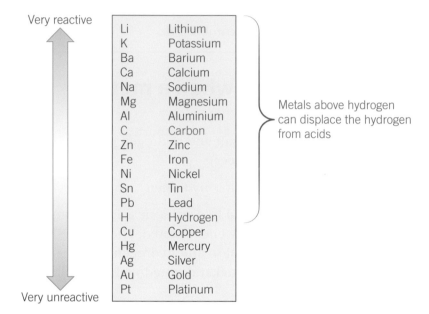

Very reactive

Li	Lithium
K	Potassium
Ba	Barium
Ca	Calcium
Na	Sodium
Mg	Magnesium
Al	Aluminium
C	Carbon
Zn	Zinc
Fe	Iron
Ni	Nickel
Sn	Tin
Pb	Lead
H	Hydrogen
Cu	Copper
Hg	Mercury
Ag	Silver
Au	Gold
Pt	Platinum

Very unreactive

Metals above hydrogen can displace the hydrogen from acids

1. Why does magnesium replace hydrogen when it reacts with hydrochloric acid?

2. Would pure silver jewelry react with an acid?

3. Predict how the reaction between lead and hydrochloric acid would compare to the reaction between magnesium and hydrochloric acid. Describe how your observations of the two reactions would differ.

4. Look back at the question earlier in the chapter about sacrificial anodes. What do you notice about their position in the reactivity series?

 Experiment

Reaction of a metal with an acid

 Safety

- Hydrochloric acid is a strong, corrosive acid. Avoid contact with the skin and eyes.

- Wear eye protection.

Materials

- 1.0 mol dm^{-3} hydrochloric acid

- 5 cm^3 measuring cylinder or graduated pipette

- Test tubes and test tube rack

- Rubber test tube stopper

- Magnesium, aluminium, zinc, iron, nickel and copper metal strips

Method

1. Place six test tubes in a rack.

2. Add 5 cm^3 of 1.0 mol dm^{-3} hydrochloric acid to each test tube using a 5 cm^3 measuring cylinder or graduated pipette.

3. Add one or two small pieces of one of the following metals to one of the test tubes: magnesium, aluminium, zinc, iron, nickel or copper.

4. If a reaction is observed, put a rubber stopper at the top of the test tube to stop any gas from escaping. When the pressure starts to build, remove the stopper and place a lit wooden splint at the mouth of the test tube. Record your observations.

5. Repeat for the other metals. Use a different test tube each time

Questions

1. Discuss the differences in the level of reactivity of each metal.

2. Order the reactions from vigorous to unreactive based on your qualitative observations.

What role do electrons play in redox reactions?

REDOX

The reactivity of an element is an indication of the ease with which an atom gains or loses one or more electrons. The effect of this reactivity is discussed throughout this book.

 Experiment

This experiment allows you to observe how rapidly electron transfer can occur in a redox reaction.

 Safety

- Dispose of waste copper(II) sulfate as instructed by your teacher.

- Wear safety glasses.

Materials

- 0.2 mol dm^{-3} copper(II) sulfate solution

- Zinc metal

- 200 cm^3 beaker

- Tweezers

Method

1. Half fill a 200 cm^3 beaker with 1.0 mol dm^{-3} copper(II) sulfate solution.

2. Take a piece of zinc metal (about 5 cm × 2 cm) from its container.

3. Hold the zinc with tweezers and dip it halfway into the copper solution for a few seconds and record your observations.

4. Place the zinc into the solution and leave for 15 minutes.

Questions

1. Describe the appearance of a fresh piece of zinc metal.

2. Describe what happens when you dip the zinc metal into copper(II) sulfate solution. Explain the type of reaction you think has occurred.

3. How does the zinc's appearance change after it has been in the solution for 15 minutes?

4. Write a balanced chemical equation to describe the reaction.

▲ The reaction between zinc and copper(II) sulfate is very rapid. How can you deduce a reaction has occurred?

In accordance with the reactivity series, in the previous experiment the copper(II) ion has been displaced by the more reactive zinc metal. The ionic equation for this reaction is more descriptive and allows us to see what is happening more clearly. An ionic equation removes the spectator ions from the chemical equation. Spectator ions are ions that are not directly involved in the reaction.

Worked example: Half-equations

Question

What are the half equations for the reaction between zinc and copper(II) sulfate?

Answer

Write the balanced chemical equation for the reaction:

$$Zn(s) + CuSO_4(aq) \rightarrow ZnSO_4(aq) + Cu(s)$$

Then, write the ionic equation by separating the soluble ionic compounds into their constituent ions:

$$Zn(s) + Cu^{2+}(aq) + SO_4^{2-}(aq) \rightarrow Zn^{2+}(aq) + SO_4^{2-}(aq) + Cu(s)$$

Find the net ionic equation by removing the spectator ions from the equation. These are the ions that are found in the same form on both sides of the chemical equation.

In this case, the sulfate ion remains unchanged throughout the reaction, so the ionic equation is:

$$Zn(s) + Cu^{2+}(aq) \rightarrow Zn^{2+}(aq) + Cu(s)$$

Remember that the reactant and product sides of the equation must be balanced:

- Check that there are the same number of atoms of each of the elements on each side of the equation.

- Check that the net charge is the same on each side of the equation. In this case the reactant side has a net charge of 2+, and the product side has a net charge of 2+.

Oxidation occurs when electrons are lost by a species or transferred to the other species in the reaction. We can identify the species losing or gaining electrons by separating the net ionic equation into two half-equations (see Chapter 10, Movement).

Then we can find balanced half equations for the reaction.

- Identify the changes to each element:

$$Zn(s) \rightarrow Zn^{2+}(aq) \qquad\qquad Cu^{2+}(aq) \rightarrow Cu(s)$$

- Balance the charges by adding electrons to one side:

$Zn(s) \rightarrow Zn^{2+}(aq) + 2e^-$

In this half-equation, electrons are being lost from zinc.

The zinc atom has been oxidized.

$Cu^{2+}(aq) + 2e^- \rightarrow Cu(s)$

In this half-equation, electrons are being gained by the copper(II) ion.

The copper(II) ion has been reduced.

Notice that the electrons are added to opposite sides of the half equations. The two half-equations are interacting: the electrons that are lost from one species are transferred to the other species. The process of oxidation cannot occur without reduction.

O	**Oxidation**
I	**Is**
L	**Loss (of electrons)**
R	**Reduction**
I	**Is**
G	**Gain (of electrons)**

▲ The mnemonic OILRIG may help you remember the difference between oxidation and reduction

1. For each of the following ionic equations:

   a) Balance the equation so the net charge is the same on each side of the equation

   b) Construct the two half-equations from the ionic equation

   c) Balance the charges for each half-equation by adding electrons

   d) Identify the species that is being oxidized and the species being reduced.

$$Cu^{2+} + Mg \rightarrow Cu + Mg^{2+}$$

$$Al + Fe^{3+} \rightarrow Al^{3+} + Fe$$

$$Ag^+ + Cd \rightarrow Cd^{2+} + Ag$$

$$Sn + Pb^{2+} \rightarrow Sn^{4+} + Pb$$

2. Construct the half-equations for the rusting of iron.

Summative assessment

Statement of inquiry:

The interactions between substances can sometimes be understood and predicted by examining the underlying processes.

Introduction

In this summative assessment you will start by analyzing the interaction between different chemical systems and how these interactions can be predicted. You will then design an experiment to examine the differences in the reactivity of metals and analyze the methodology used by a student to investigate how the rate of production of hydrogen gas changes with different metals. Finally, you will learn about a creative approach to solving the global problem of corrosion by means of anti-corrosion coatings.

 Types of reactions

1. Combustion reactions are an example of an oxidation reaction.

   a) What is a combustion reaction? Illustrate your answer with a word equation or balanced chemical equation. [2]

   b) Give three examples of combustion reactions that play a role in your daily life. [3]

   c) Explain the difference between complete and incomplete combustion reactions. [4]

2. A student performed a series of reactions between a metal and solution and recorded whether or not there was a reaction. Consider the results and answer the following questions.

	Silver nitrate AgNO$_3$	Iron(III) chloride FeCl$_3$	Magnesium iodide MgI$_2$	Copper(II) nitrate Cu(NO$_3$)$_2$	Aluminium chloride AlCl$_3$	Potassium nitrate KNO$_3$
Silver Ag		✗	✗	✗	✗	✗
Iron Fe	✓		✗	✓	✗	✗
Magnesium Mg	✓	✓		✓	✓	✗
Copper Cu	✓	✗	✗		✗	✗
Aluminium Al	✓	✓	✗	✓		✗
Potassium K	✓	✓	✓	✓	✓	

a) Identify the most reactive metal. [2]

b) Iron is less reactive than aluminium. Provide evidence to justify this statement. [2]

c) State which is the least reactive metal and give reasons for your choice. [2]

d) List the metals in order of most reactive to least reactive. [3]

3. The single replacement reactions in question 2 are examples of a redox reaction. For each of the reactions listed below:

   a) write the balanced chemical equation for the reaction [8]

   b) write the two half-equations for the reaction [8]

   c) identify the species being oxidized and the species being reduced. [8]

      i) Silver nitrate and magnesium

      ii) Aluminium and copper(II) nitrate

      iii) Potassium and magnesium iodide

      iv) Iron and copper(II) nitrate (hint: iron(III) is one of the products)

Identifying the reactivity of a metal from experimental data

The reaction between a metal and an acid produces a soluble salt and hydrogen gas. The rate of production of hydrogen gas is a consequence of the reactivity of the metal present during the reaction. To ensure that the reaction rate is a reflection of the reactivity of the metal, you will need to select the dependent and independent variable, and design a method that controls all other variables.

4. Design an experiment to investigate the effect the reactivity of a metal has on the initial rate of production of hydrogen gas. Use the following points for guidance. [5]

Equipment

- Freshly polished pieces of metal of equal surface area (2 cm × 2 cm) including copper, zinc, tin, magnesium, aluminium, zinc and iron.

- 1.0 mol dm^{-3} hydrochloric acid

- 250 cm^3 Büchner flask

- Glass gas syringe

- Rubber bung

- Stopwatch

> **Method**
>
> - Identify the independent, dependent and control variables.
> - Describe the quantitative and qualitative observations you plan to make.

5. Explain the importance of the control variables in this experiment. [3]

6. Explain how the method you have designed enables you to examine the initial rate of reaction and hence make a comparison between the reactivity of the metals. [2]

 ## Analysis and evaluation

Systematic errors are associated with flaws in the experimental design or instrumentation used while random errors are a result of uncontrolled variables.

A student performed a series of experiments to investigate how the reactivity of a metal affects the rate of formation of hydrogen gas when it is reacted with a strong acid. The method followed is outlined below.

> **Materials**
>
> - Zinc powder, magnesium ribbon and copper sheet
> - 1.0 mol dm^{-3} hydrochloric acid
> - Electronic balance
> - Glassware
> - Stopwatch
> - Cotton wool
>
> **Method**
>
> 1. Collect samples of zinc powder, magnesium ribbon and copper sheet.
>
> 2. Weigh out 5 g samples of each metal and place them into individual 100 cm³ conical flasks.
>
> 3. Using a beaker, measure 20 cm³ of 1.0 mol dm^{-3} hydrochloric acid. Transfer this into one of the conical flasks containing metal.
>
> 4. Place a cotton wool ball into the mouth of the conical flask and place the flask on the electronic balance.
>
> 5. Start the stopwatch and record the initial mass.
>
> 6. Every 20 s record the mass of the flask until the reaction is complete.
>
> 7. Repeat steps 3 to 6 for each type of metal.
>
> 8. Record your results in a data table.

7. Identify any systematic errors or random errors in the method designed by the student. [4]

8. Evaluate the method and identify any steps that require modification. Explain your reasoning. [4]

9. Explain how you could improve the method or extend the scope of the investigation. [4]

Environmentally friendly anti-corrosion coatings

The following text is from the website of the United States Environmental Protection Agency. It outlines a project to investigate alternative anti-corrosion coatings.

Corrosion of metal structures is estimated to cost many billions of dollars annually. The most common methods of corrosion inhibition or prevention involve the application of heavy surface treatments (paints and primers) or conversion coatings using various metallics in processes that are strictly controlled and regulated due to toxicity and possible carcinogenic properties. *Luna Innovations* proposes to develop an alternate process that is capable of corrosion inhibition without the use or generation of hazardous materials, which also can be adapted to the coating of large surfaces. A new coating process has been developed based on ionic self-assembled monolayers (ISAM) that: (1) has demonstrated corrosion inhibition of aluminum alloys; (2) neither contains nor generates hazardous materials; and (3) has demonstrated practical application methods, including spraying and non-electrolytic brushing. This Phase I project will adapt the coating process to steel and other metals and alloys. The resulting coating process will allow the long-term storage of raw materials without the need for refinishing or removal prior to use. The ultrathin coating layer is compatible with all normal fabrication processes, including welding and painting. The proposed process will find application in protection of large structures such as ships, bridges, automotive components, and commercial aircraft manufacturing. It will serve as a low-cost, environmentally friendly replacement of corrosion inhibition coatings for many smaller structures and components in commercial and industrial applications.

10. Identify the problem for which the company *Luna Innovations* are aiming to develop solutions. [1]

11. Explain the risks involved in the current technology in use. [2]

12. Green chemistry aims to reduce or eliminate the use or generation of hazardous substances. List the three advantages of the ISAM process that the company is developing and discuss the advantage of the green chemistry approach. [3]

13. State four types of projects in which the process outlined could be used. [4]

14. Evaluate why this new coating process is an exciting innovation. [2]

Introduction

A scientific theory is a collection of facts, laws and tested hypotheses; it is used to explain systems and phenomena. Models can help to transform concepts that appear to be abstract into concrete concepts. They help us to make sense of complex systems.

Chemists use models to focus our thinking about the position of atoms in time and space. Models enable us to better understand microscopic and macroscopic properties. Scientists develop models of varying complexity to test scientific theories and explain processes that cannot be observed by the human eye. Models may require imagination and creativity to develop, however, they are based upon the theoretical understanding of the concept being examined. Models representative of an atom, and animal and plant cells are some of the most widely recognized models in science.

The pharmaceutical industry continues to utilize developments in computing power to advance drug design. By using the 3D structures of the protein targets of new drugs and advanced computational methods, researchers can test, interpret data and create new structures in much shorter time frames than through traditional drug testing. This has resulted in more cost-effective drug development and new drugs can be developed and tested more quickly, including the modification and, sometimes, altered usage of old drugs.

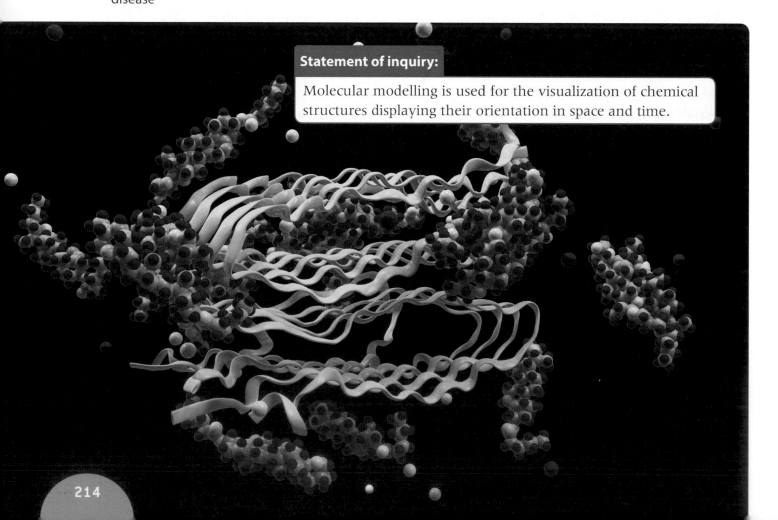

▼ Molecular model simulation of a drug (red and white) attacking the protein fibrils (ribbons) that are believed to be the cause of Alzheimer's disease

Statement of inquiry:

Molecular modelling is used for the visualization of chemical structures displaying their orientation in space and time.

FIG. 10.

Formation of MARSH-GAS.

▲ Physical models help in the visualization of different concepts in the study of chemistry. August Wilhelm von Hofmann, a German scientist, is credited with constructing the first physical molecular model in 1860. The molecule of methane was a two-dimensional representation of the molecule and the relative size of carbon and hydrogen atoms was not considered. It was not until 1874 that the three-dimensional representations of molecules accurately represented the relative sizes of atoms. These molecular models remain a valuable learning tool for people studying the molecular structure of compounds

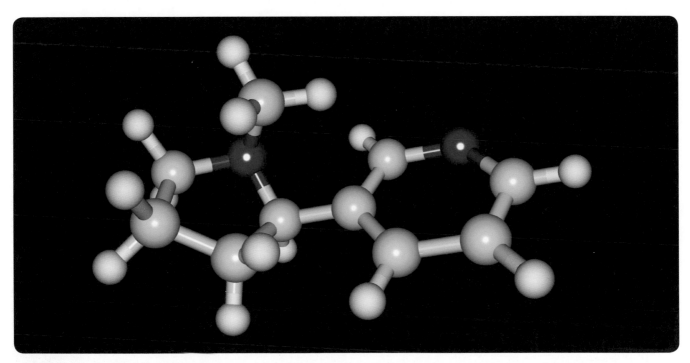

▲ Nicotine, $C_{10}H_{14}N_2$, is a naturally occurring alkaloid compound. Alkaloids are a diverse class of organic compounds which contain a nitrogen-based ring structure. Numbering over 20,000 compounds, molecular models of these compounds enable you to easily identify both similarities and differences. Some well-known examples of alkaloids include codeine (an opiate used to treat pain and a cough medication), morphine (an opiate-type pain medication), quinine (an antimalarial medication) and strychnine (a highly toxic pesticide)

Why do elements bond?

On 30 December 2015, the International Union of Pure and Applied Chemistry (IUPAC) officially declared the existence of four new elements, bringing the total number of elements to 118. In the future more elements may be discovered. The synthesis of new compounds made from elements is rapid. On average, the scientific community synthesizes a new compound every two to three seconds!

A compound is the chemically bonded combination of atoms of different elements in a fixed ratio. Atoms form bonds in order to form a stable outer shell like those found in the noble gases in group 18. The arrangement of atoms is the foundation of all that we study in chemistry. We use models to represent these combinations of atoms to better understand the chemical bonding and the geometry of these structures.

How are ions formed?

An ionic bond is defined as the electrostatic attraction between the electric charges of a cation (positive ion) and an anion (negative ion). In general, ionic compounds are formed between metals and non-metals.

Ions are formed when electrons are transferred from one atom to another (see Chapter 11, Patterns). Knowing the position of an element in the periodic table, relative to group 18 (the noble gases) helps us to predict whether an atom will lose or gain electrons and the size of its charge.

A typical example of an ionic compound is the salt sodium chloride. Sodium is found in group 1 and therefore has one valence electron. It has the electron configuration of 2,8,1 (for more on valence electrons and electron configuration see Chapter 11, Patterns).

With an atomic number of 11, sodium has two noble gases relatively close in terms of position and electron configuration.

▶ The sodium ion has two options

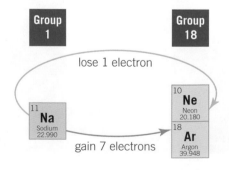

As the driving force in the formation of ions is to adopt a noble gas electron configuration, thus achieving a full outer shell of electrons, the sodium atoms has two possibilities:

● gain seven electrons to fill its outer shell and adopt the electron configuration of argon [2,8,8]

● lose one electron and adopt the electron configuration of neon [2,8].

It is easier to lose 1 electron than gain 7. Why do you think this is? This explains why metals tend to lose electrons from their outer electron shells rather than gain them. In losing an electron, sodium is oxidized (see Chapter 8, Interaction).

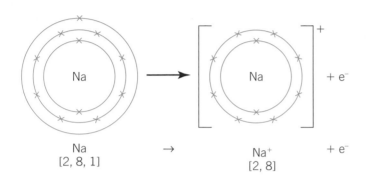

Na [2, 8, 1]   →   Na⁺ [2, 8]   + e⁻

◀ The sodium atom will lose its one valence electron, forming the sodium cation, Na⁺

1. All elements found in group 1 lose one valence electron when oxidized. What is the major difference between the electrons lost in each of these elements?

Chlorine, a non-metal, has an electron configuration of [2,8,7]. Using the same approach as for sodium, the chlorine atom could:

● gain one valence electron to achieve a full outer electron shell to be like argon [2,8,8]

● lose seven electrons to be like neon [2,8].

Non-metals gain electrons to complete the outer electron shells. By doing this they are reduced.

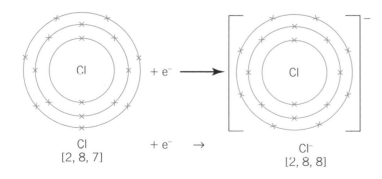

Cl [2, 8, 7]   + e⁻   →   Cl⁻ [2, 8, 8]

◀ The chlorine atom will gain one valence electron, forming the chlorine anion, Cl⁻

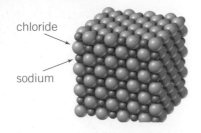

chloride

sodium

▲ The crystal lattice of sodium chloride is held together by the strong electrostatic interactions between oppositely charged ions. Oppositely charged ions surround each other in the crystal lattice structure

The electron lost by sodium is gained by chlorine. Electron transfer creates these charged ions which now experience an electrostatic attraction to one another. However, the sodium and chloride ions don't stay in pairs—they form a regular lattice of alternating positive and negative ions, and the ionic compound sodium chloride is formed.

▲ A highly magnified image of a sodium chloride crystal reveals the geometric patterns present in a crystalline lattice structure

GENERAL

How can we predict the charge on an ion?

To predict the charge of an ion, you need to examine the balance between the number of positively charged protons in the nucleus of the atom and the number of negatively charged electrons in the atomic orbitals surrounding the nucleus.

The atomic number of an element is the number of positively charged protons found inside the nucleus of the atom. The mass number is the total number of protons and neutrons found in the nucleus of the atom. The mass of an electron is so small, it is effectively ignored when determining the mass of an atom.

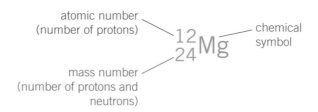

atomic number (number of protons) — $^{24}_{12}\text{Mg}$ — chemical symbol

mass number (number of protons and neutrons)

▶ Describing an element

Magnesium has an atomic number of 12 and a mass number of 24. This means that the element has 12 protons and therefore 12 electrons, with an electron configuration of [2,8,2]. It also has 12 neutrons (24 – 12). This metal atom will lose two electrons to achieve a full outer electron shell, leaving the atom with ten electrons and a new electron configuration of [2,8]. It now has the same electron configuration as neon. When two atoms or ions have the same electron configuration, they are said to be isoelectronic with each other.

1. Give the names, symbols and charges of three cations and three anions that are isoelectronic with the noble gas neon [2,8].

Having lost two negatively charged electrons, the magnesium ion no longer has the same number of electrons and protons. This charge imbalance results in the magnesium ion having a 2+ charge.

Number of:	Magnesium atom, Mg	Magnesium ion, Mg^{2+}
protons	12	12
neutrons	12	12
electrons	12	10
net charge	0	2+

▲ Comparison of magnesium atom and ion

1. Copy and complete the following table.

Element name	Elemental symbol	Atomic number	Mass number	Number of protons	Number of neutrons	Number of electrons	Charge	Symbol of ion and type of ion (cation/anion)
lithium	Li	3	7			2		
	Al		27	13			3+	Al^{3+}/cation
calcium				20	20		2+	
			56			23		Fe^{3+}/cation
	Si	14				10	4+	
oxygen			16		8			O^{2-}/anion
	N		14	7			3–	
chlorine	Cl	17			18			Cl^-/anion

How do chemists model ionic compounds?

The chemical formula of an ionic compound tells you the elements present in the compound and the lowest whole number ratio of the atoms present. For example, sodium chloride has the chemical formula NaCl. The formula indicates that there is one sodium ion for every chloride ion.

The chemical formula does not give any indication of the actual structure of an ionic compound. Models help us to visualize what we cannot see. Under normal conditions, ionic compounds are solids with a crystalline lattice structure. This three-dimensional structure is made of repeating units—the positive and negative ions described by the chemical formula.

▶ This three-dimensional model of the ionic compound sodium chloride clearly shows the orientation in space of the ions in relation to one another. Focus on a single charged ion and you will notice how the next nearest neighbor in all directions is an ion of the opposite charge

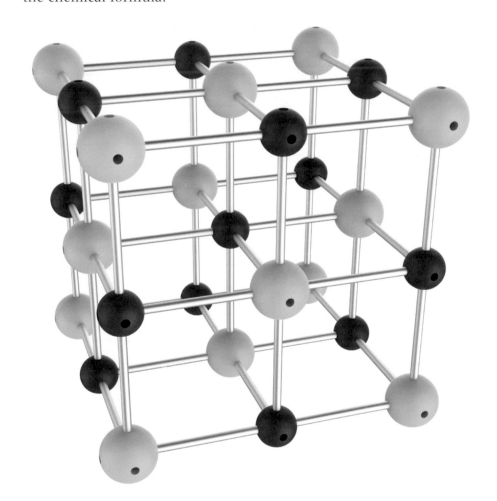

Using models of compounds helps to bring the structures to life. They help us to understand how structure influences properties such as melting and boiling points, volatility, electrical conductivity and solubility in solvents.

Why do ionic compounds have high melting and boiling points?

The electrostatic attraction between oppositely charged ions within the lattice structure are strong. The structure for the compound sodium chloride is said to be a 6:6 coordinate structure. This means that each positively charged sodium ion is bonded to six negatively charged chloride ions, and vice versa. Examine the model of sodium chloride and confirm this for yourself.

It requires a lot of energy to break these very strong forces of attraction between the ions. Sodium chloride has a melting point of 801°C and a boiling point of 1,413°C. Models of sodium chloride and other ionic crystal structures enable us to better understand and explain the concepts of melting and boiling points.

ATL Creative thinking skills

Apply existing knowledge to generate new ideas, products or processes

We have established that the strength of an ionic bond is due to the close proximity of oppositely charged ions. In small groups, discuss the following.

1. Magnesium oxide is made up of Mg^{2+} ions and O^{2-} ions. How does the amount of charge present on the ions in an ionic compound affect the melting and boiling points? Explain your reasoning.

2. The strength of an ionic bond is also dependent on the distance between the two ionic centres. How are the melting and boiling points are affected by the ionic radius of the interacting ions? Sodium chloride has a melting point of 801°C and a boiling point of 1,413°C. Cesium has a larger ionic radius than sodium, and iodine has a larger ionic radius than chlorine. How do the melting and boiling points of sodium chloride and cesium iodide compare? Explain your reasoning.

Why are ionic compounds brittle?

The crystalline lattice structure of an ionic compound is hard and dense because the ions are closely packed together. Because they are arranged so that only oppositely charged ions are touching one another, the structure is extremely stable. The model enables you to visualize this. If the crystalline structure was placed under stress, such as being hit with a hammer, this could result in a shift in the planes of ions. If ions of the same charge came into contact, the repulsive forces would cause the crystal to shatter.

Smashing rock salt

You can try this at home. Take a piece of rock salt—this is sodium chloride. Take the flat surface of a knife and press down on the crystal. What do you observe?

▲ 1 mole of rock salt has a mass of 58 g. 1 mole contains 6.02×10^{23} sodium ions and the same number of chloride ions. Consider how many individual ions you are observing!

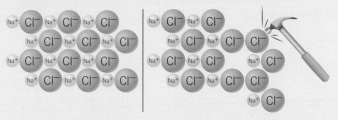

▲ Stress applied to a crystal of sodium chloride results in like-charged ions moving adjacent to one another and subsequently repelling each other. This causes the crystal to shatter. What happens when you move like poles of two bar magnets close to one another?

GENERAL

What affects the solubility of ionic compounds in water?

An ionic compound may dissolve in a polar solvent such as water. It won't dissolve, however, in a non-polar solvent such as hexane. The rule of thumb (see Chapter 7, Function) is: like dissolves like.

Water molecules are polar because their shared electrons are unevenly distributed. The hydrogen part has a partial positive charge and the oxygen part has a partial negative charge.

When the attraction between the negative (oxygen) part of the water molecule and the sodium cation is greater than the electrostatic attraction between the sodium and chloride ions within the lattice structure, the sodium ions leave the lattice structure and are enveloped by water molecules.

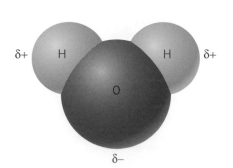

▲ Water molecules are polar

Similarly, the chloride anions are strongly attracted to the positive (hydrogen) part of the water molecule, so they leave the lattice structure and are surrounded by the water molecules.

The combination of these effects results in the crystal beginning to dissolve. The process is known as solvation.

For ionic compounds where the charges on the ions are greater or the ions are in closer proximity, the bonds between the ions are stronger, so the ions are less attracted to the polar water molecules. As a result, these compounds are less soluble.

1. Do you expect magnesium oxide, MgO, to be more soluble or less soluble than sodium chloride? Explain your reasoning.

2. Scientists use three-dimensional visualizations or models to enable a clear understanding of the concept being examined. How do other professions use modelling or simulations to convey meaning?

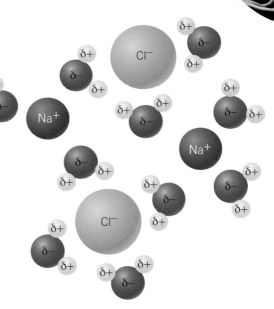

▲ Solvation is the process where an ionic solid is dissolved by a polar solvent such as water

GENERAL

What is a supersaturated solution?

A saturated solution is one in which no more solute can be dissolved in the solvent at a given temperature. By increasing the temperature of the solvent, we can increase the level of solubility of the solute, creating a supersaturated solution. A supersaturated solution has a solubility that exceeds the level possible at room temperature, for a specific solute.

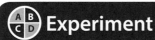 **Experiment**

Investigating a supersaturated solution

 Safety

● Wear safety glasses.

Materials

● 440 g sodium hydrogen carbonate (also known as baking soda or sodium bicarbonate)

● 500 cm³ white vinegar

● 100 cm³ distilled water

● 1 dm³ beaker

● Saucepan or another 1 dm³ beaker

● Bunsen burner or other heat source

● Wooden splint or chopstick

Method

1. Place the baking soda into a stainless steel saucepan or 1 dm³ beaker.

2. Pour the vinegar onto the baking soda and mix well. Leave it to stand for 10 minutes or until the effervescence stops.

3. Warm the mixture over a gentle heat until the solution becomes clear.

4. Remove the heat source, pour most of the solution into a clean beaker, cover it and leave it to cool to room temperature. Do not let any dust or foreign matter fall into the mixture.

5. Cover the end of the wooden splint or chopstick with crystals from the saucepan or original beaker. A small amount is sufficient. These are called seed crystals.

6. Remove the cover from the beaker and place the tip of the splint into the centre of the beaker.

Questions

1. What did you observe when you placed the seed crystal into the saturated solution?

2. Draw conclusions on what is happening to the solubility of the salt as the temperature of the solution decreases. Support your discussion with scientific reasoning.

GENERAL

How is the formula of an ionic compound constructed?

The octet rule states that to attain a noble gas core electron configuration, elements do one of the following:

- lose electrons and undergo oxidation

- gain electrons and undergo reduction

- share electrons.

When you analyze the formula of an ionic compound, the overall charge for the compound is zero. The number of positive charges from the cation and the number of negative charges from the anion are balanced. This enables you to construct the formula of an ionic compound. The approach for metals in groups 1,2 and 13 is slightly different to that for the transition metals found in groups 3 to 12, so we will look at each method separately.

Thinking in context

Assigning valence electrons of an element using the periodic table

The position of an element in the periodic table tells us:

- whether it is classified as a metal or a non-metal

- what group it belongs to and from this some of the properties common to this group

- how many valence electrons it has.

For example, sodium is found in group 1—it has one valence electron. Calcium is in group 2 and has two valence electrons. However, aluminum is in group 13. For elements in a group number greater than 10, simply drop the "1" to find the number of valence electrons. So, aluminum has three valence electrons and silicon in group 14 has four valence electrons.

Worked example: Formulas of group 1 and 2 metal halides

Question

Find the formula of metal halides from groups 1 and 2— potassium iodide and magnesium iodide.

Answer

First, identify the metal and non-metal elements, their symbol and their charge.

In the formula for an ionic compound, the symbol for metal, or positively charged cation, appears first and the symbol for the non-metal or negatively charged anion second.

Element	Symbol	Group	Electron configuration	Ion
potassium	K	1	2,8,8,1	K^+
magnesium	Mg	2	2,8,2	Mg^{2+}
iodine	I	17	2,8,8,7	I^-

Then balance the charges. Ask yourself if the charges on the cation and anion are the same and therefore balance each other.

If the charge on the cation and anion are identical in magnitude, the formula of the compound is a 1:1 ratio of the elements. Overall, the compound has no charge.

For example, potassium iodide contains K^+ and I^- ions. As the charges balance, the formula for the compound potassium iodide is KI. (Remember, ionic compounds are always expressed as the lowest whole number ratio of the atoms in the compound.)

If the charges on the individual ions do not cancel out one another, as with Mg^{2+} and I^-, we can use the criss-cross method to work out how to balance the charges.

To use the criss-cross method, swap the size of the charge and ignore its sign. This shows that the formula of magnesium iodide is MgI_2

$$Mg^{2+}I^- \rightarrow MgI_2$$

Therefore, the formula for the ionic compound magnesium iodide is MgI_2. The magnesium ion has a charge of 2+ and there are two iodide ions, each with a 1– charge, and so the formula has a net charge of zero.

Finally, we can determine the nomenclature. For a simple binary ionic compound, the name of the cation is the unmodified element name. For the anion, the suffix is changed from iod-ine to iod-ide.

1. For each of the following pairs of elements, deduce the formula and name.

   a) Sodium and fluorine

   b) Lithium and oxygen

   c) Potassium and nitrogen

   d) Calcium and chlorine

   e) Magnesium and sulfur

   f) Strontium and phosphorus

   g) Aluminium and bromine

   h) Aluminium and oxygen

   i) Aluminium and nitrogen

Transition metal individual elements are capable of existing as many different cations.

Sc	Ti	V	Cr	Mn	Fe	Co	Ni	Cu	Zn
		1+	1+	1+	1+	1+	1+	1+	
2+	2+	2+	2+	2+	2+	2+	2+	2+	2+
3+	3+	3+	3+	3+	3+	3+	3+	3+	
	4+	4+	4+	4+	4+	4+	4+		
		5+	5+	5+					
			6+	6+	6+				
				7+					

▶ The transition metals can form a wide range of ions when reacting with different species. Which charged ion can all transition metals form?

For example, manganese can be found in compounds as Mn^+, Mn^{2+}, Mn^{3+}, Mn^{4+}, Mn^{5+}, Mn^{6+} and Mn^{7+}. The most common ions are the manganese(II) and manganese(VII) ions. Notice how Roman numerals are used to represent the size of the charge on a transition metal ion.

Worked example: Formulas of transition metal halides

Question

Find the formula of transition metal halides—iron(III) chloride and cobalt(II) chloride.

Answer

Identify the metal and non-metal elements, as well as the symbol and charge of each.

"Iron(III)" and "cobalt(II)" tells you that the cations are Fe^{3+} and Co^{2+} and the anion of chlorine is Cl^-.

Balance the charges using the criss-cross method.

$$Fe^{3+}Cl^- \rightarrow FeCl_3$$

$$Co^{2+}Cl^- \rightarrow CoCl_2$$

So, the formulas of iron(III) chloride and cobalt(II) chloride are $FeCl_3$ and $CoCl_2$.

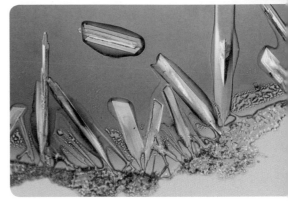

▲ Hydrated iron(III) chloride crystals magnified

What is a polyatomic ion?

GENERAL

A polyatomic ion is sometimes referred to as a molecular ion. It is composed of two or more covalently bonded atoms (you will find out more about this later in the chapter) with a charge. Polyatomic ions are commonly seen in the formulae of acids, salts and transition metal complexes. When combining them in ionic compounds, the same rules for the formation of a compound can be applied.

Some common polyatomic ions are:

ammonium ion	NH_4^+	nitrate ion	NO_3^-
carbonate ion	CO_3^{2-}	peroxide ion	O_2^{2-}
ethanedioate ion	$C_2O_4^{2-}$	phosphate ion	PO_4^{3-}
hydrogencarbonate ion	HCO_3^-	phosphonate ion	PO_3^{2-}
hydroxide ion	OH^-	sulfite ion	SO_3^{2-}
nitrite ion	NO_2^-	sulfate ion	SO_4^{2-}

Parentheses (brackets) are important when a compound contains multiple polyatomic ions. The use of parentheses is illustrated in the compound on iron(III) hydroxide. Three hydroxide ions will bond with the iron(III) ion. Parentheses here signify that there are three OH^- ions. If the parentheses were omitted, what would the formula look like?

$$Fe^{3+}OH^- \rightarrow Fe(OH)_3$$

▲ Using the criss-cross method to find the formula of iron(III) hydroxide

In the case of ammonium phosphate, parentheses are also required. How would this formula differ with the parentheses omitted?

$$NH_4^+PO_4^{3-} \rightarrow (NH_4)_3PO_4$$

▲ Using the criss-cross method to find the formula of ammonium phosphate

1. For each of the following pairs of elements, deduce the formula and name.

   a) Iron(III) and nitrate ion

   b) Vanadium(III) and sulfate ion

   c) Manganese(VI) and oxide ion (Hint: ionic compounds are the lowest whole number ratio)

   d) Copper(I) and carbonate ion

   e) Copper(II) and sulfate ion

   f) Ammonium ion and nitrite ion

   g) Ammonium ion and sulfite ion

   h) Nickel(IV) and oxide ion

   i) Manganese(VII) and sulfate ion

   j) Titanium(II) and hydrogencarbonate ion

How do the chemical formulae of covalent and ionic compounds differ?

GENERAL

We have established that the chemical formula of an ionic compound represents the lowest whole number ratio of atoms in a compound. The ionic compound aluminium oxide, Al_2O_3, is a giant crystalline lattice of millions of aluminium cations and oxygen anions in the fixed ratio 2:3.

▲ Aluminium ore or bauxite is the impure form of aluminium oxide. When pure aluminium metal undergoes the process of oxidation, a protective layer is formed that prevents further oxidation of the metal. Aluminium oxide has a hexagonal close-packed structure

A covalent bond is defined as an electrostatic attraction between a shared pair of electrons and the positively charged nuclei. As with ionic compounds, atoms of elements involved in covalent bonding also want to attain a noble gas core electron configuration. Rather than transferring electrons from one atom to another to form charged ions, non-metal atoms share electrons to attain a full outer electron shell.

Examples of covalent-bonded compounds are some of the most well-known substances. When asked for the chemical formula of water, carbon dioxide, oxygen gas or nitrogen gas, most people would be able to give an accurate answer. These formulae are frequently seen in the media, in advertising and articles written for the general public.

The formula for simple covalent compounds can be determined by following these steps:

- work out the number of valence electrons for each element
- create a Lewis symbol for each element
- apply the octet rule to create Lewis structures for the compound.

For example, the elements of carbon, nitrogen and sulfur are found in groups 14, 15 and 16 respectively. How many valence electrons does each atom have?

- Carbon is in group 14, so by dropping the "1" from 14, we know that it has four valence electrons.
- Nitrogen is in group 15, so it has five valence electrons.
- Sulfur is in group 16, so it has six valance electrons.

A Lewis symbol for an element shows only the outermost, valence electrons involved in bonding. They are placed around the element symbol in the order shown to the left.

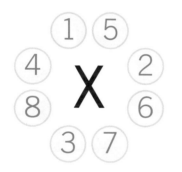

▲ Order of placing the valence electrons in a Lewis symbol

The valence electrons can be represented by either • or ×. Following these rules, the diagram below shows the Lewis symbols for carbon, nitrogen and sulfur.

1. For each of the following elements determine the number of valence electrons and its Lewis symbol.

hydrogen	phosphorus	fluorine	chlorine
boron	silicon	oxygen	

▲ Lewis symbols for carbon, nitrogen, and sulfur

A Lewis electron dot structure is a visual representative model of a molecule formed by a covalent bond between two or more non-metal atoms. Lewis symbols provide us with the information to understand how non-metal atoms combine.

Many of the common gases we are familiar with are diatomic molecules. This means that the molecule is made up of two identical atoms chemically bonded to each other. Examples of diatomic molecules include hydrogen, fluorine, chlorine, oxygen and nitrogen gas.

Worked example: Lewis electron dot structures

Question

Work out the Lewis electron dot structures for hydrogen, fluorine, and oxygen.

Answer

Hydrogen, H$_2$

Hydrogen is found in group 1 and has one valence electron. It already has one electron, so it only needs to acquire one electron to fill its outer shell to attain the noble gas electron configuration of helium. This can be achieved by two hydrogen atoms sharing one valence electron.

$$\dot{H} + \dot{H} \rightarrow H\!:\!H \quad \text{or} \quad H-H$$

> a **single bond** has a shared pair of electrons and is represented by a line —

▲ Lewis electron dot structure for hydrogen

Fluorine, F$_2$

Fluorine is found in group 17 and has seven valence electrons. Fluorine [2,7] needs to acquire one electron to fill its outer electron shell and attain the noble gas electron configuration of neon [2,8], thus satisfying the octet rule. This can be achieved by two fluorine atoms sharing one valence electron.

$$\ddot{\underset{..}{F}}\!\cdot + \cdot\ddot{\underset{..}{F}}\!: \rightarrow :\ddot{\underset{..}{F}}\!:\!\ddot{\underset{..}{F}}\!: \quad \text{or} \quad :\ddot{\underset{..}{F}}\!-\!\ddot{\underset{..}{F}}\!:$$

▲ Lewis electron dot structure for fluorine

▲ Fluorine is one of the most reactive elements in existence. When in its combined form, compounds that contain fluorine are relatively harmless. Fluoridated mains water and toothpaste help prevent tooth decay by strengthening the enamel coating of teeth. Unconfined, it is so reactive that research chemists rarely work with this gas. Many chemists were injured before Henri Moissan first successfully isolated fluorine in 1886

Oxygen, O$_2$

Oxygen [2,6] is found in group 16 and has six valence electrons. It needs to acquire two electrons to fill its outer electron shell and attain the noble gas electron configuration of neon, thus satisfying the octet rule.

$$\ddot{\underset{..}{O}}\!: + \cdot\ddot{\underset{.}{O}}\!: \rightarrow :\ddot{\underset{..}{O}}\!:\!:\!\ddot{\underset{..}{O}}\!: \quad \text{or} \quad :\ddot{O}=\ddot{O}:$$

> a **double bond** has two shared pairs of electrons and is represented by two lines =

▲ Lewis electron dot structure for oxygen

In this example the two oxygen atoms are sharing a total of two pairs of electrons or four electrons. This results in a double bond being formed between the two oxygen atoms. Some atoms form a triple bond in which there are three shared pairs of electrons; these are represented by three lines ≡.

Build structural models

Molecular model kits are a good way of understanding and visualizing the structure of simple compounds. They help you to determine the type of bonding between atoms, the three-dimensional nature of the compound and they can be used to confirm any Lewis structures you have developed.

 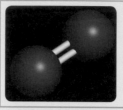 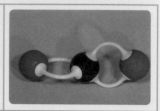

▲ Molecular models are three-dimensional representations of molecular compounds. The models visualize an atom's orientation in space relative to the other atoms in a molecule

For each of the listed compounds follow these steps.

1. Determine the number of valence electrons for each atom in the compound from its electron configuration.

2. Draw the Lewis symbol for each element.

3. Predict the number of electron pairs to be shared in the molecule.

4. Draw the Lewis structure for the molecule.

5. Using a molecular model kit, build the physical molecular model for the compound and compare it with your Lewis structure. Can you confirm the theoretical representation you developed?

6. Identify whether the molecule has single, double or triple bonds.

7. Describe the shape of each molecule.

a) chlorine, Cl_2	d) methane, CH_4	g) carbon monoxide, CO
b) bromine, Br_2	e) nitrogen, N_2	h) ammonia, NH_3
c) hydrogen fluoride, HF	f) carbon dioxide, CO_2	

What are the rules for naming covalent compounds?

The IUPAC rules for the naming of covalent compounds are relatively straightforward. Unlike ionic compounds, the molecular formula of a covalent compound describes the actual number of individual atoms in a molecule of the compound.

A molecule of the gas, carbon dioxide, CO_2, has one carbon atom and two oxygen atoms chemically bonded together. In contrast, the crystalline lattice structure of sodium fluoride, NaF, is made up of millions of sodium ions and fluoride ions chemically bonded in the fixed ratio of 1:1.

1	mono
2	di
3	tri
4	tetra
5	penta

▲ Numerical multipliers represent the number of individual atoms

The nomenclature rules are listed below.

- For a single atom in the first position, never use the numerical multiplier "mono-". Only use the name of the element.

- For multiple atoms in the first position, use a numerical multiplier in the prefix position before the name of the element.

- For single or multiple atoms in the second position, use a numerical multiplier in the prefix position before the name of the element.

- The suffix of the element in the second position changes to "-ide". For example, chlor**ine** becomes chlor**ide**. This is the same as in ionic compounds.

When two vowels come together, one vowel is deleted—it is monoxide not mono-oxide, for example.

Worked example: Naming covalent compounds

Question

For the following molecules identify the individual elements and the number of atoms of each element, and apply the nomenclature rules:

- CO
- CO_2
- P_2O_5

Answer

CO

1 Identify the individual elements:
   The elements present are carbon and oxygen.

2 Identify the number of atoms of each element:
   There is one carbon atom and one oxygen atom.

3 Apply the nomenclature rules:
   *Carbon **mon**oxide*

CO₂

1 Identify the individual elements:
   The elements present are carbon and oxygen.

2 Identify the number of atoms of each element:
   There is one carbon atom and two oxygen atoms.

3 Apply the nomenclature rules:
   *Carbon **di**oxide*

P₂O₅

1 Identify the individual elements.
   The elements present are phosphorus and oxygen.

2 Identify the number of atoms of each element and assign the numerical multiplier:
   There is two phosphorus atoms and five oxygen atoms.

3 Apply the nomenclature rules:
   ***Di**phosphorus **pent**oxide*

1. For each of the following formulae work out the IUPAC names.

a) BF_3 b) CH_4 c) SO_3 d) NF_3 e) N_2O_4
f) Cl_2O_5 g) PCl_3 h) H_2O_2 i) BCl_3 j) OCl_2

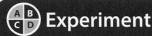 **Experiment**

Testing for conductivity

Conductivity is a measure of the presence of dissolved ions in an aqueous solution. From these observations you can draw conclusions about the composition of the solid and the types of bonding present in the compound.

⚠ Safety

- Wear safety glasses.
- Dispose of solutions appropriately.

Materials

- 0.1 mol dm^{-3} solution of sodium chloride
- 0.1 mol dm^{-3} solution of sugar
- 0.1 mol dm^{-3} solution of potassium chloride
- 0.1 mol dm^{-3} solution of hydrochloric acid
- 0.1 mol dm^{-3} solution of sulfuric acid
- 0.1 mol dm^{-3} solution of sodium hydroxide
- 0.1 mol dm^{-3} solution of ammonia
- 0.1 mol dm^{-3} solution of ammonium hydroxide
- Tap water
- Distilled water
- 100 cm^3 measuring cylinder
- 100 cm^3 beakers
- Labels
- Paper towels

- Data-logger conductivity meter or conductivity apparatus made from 9 V battery, LED, connecting wires and two inert electrodes

Method

1. Transfer 50 cm^3 of each solution into a 100 cm^3 beaker. Label the beaker.

2. Place the conductivity probe into the solution. Depending on your apparatus, either take a reading of the level of conductivity on the meter or observe the brightness of the LED. Record your results in an appropriate data table.

3. After testing each solution, wash the probes with distilled water and dry with paper towels.

Questions

1. Classify each solution as either a conductor or a non-conductor.

2. For each of the conductors, classify them as poor, good and very good conductors.

3. Find out each compound's chemical formula and classify it as ionic or covalent. Are the results with respect to conductivity what you predicted given your knowledge of the properties of these compounds?

How do metal ions bond?

Metallic bonding results from the electrostatic attraction between a lattice of positive ions and a "sea" of delocalized electrons. The manner in which the metal ions pack together has been analyzed and classified, and models have been developed to help us to visualize their internal structure. The different lattice structures have names that describe their structure. The three main types of metallic lattice structure are face-centred cubic (FCC), body-centred cubic (BCC) and hexagonal close packed (HCP).

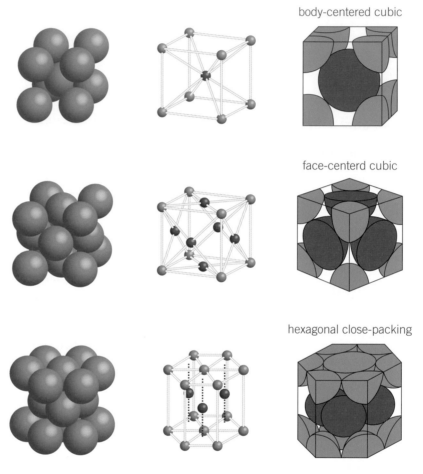

body-centered cubic

face-centerd cubic

hexagonal close-packing

▶ The unit cell structure of a metal is endlessly repeated to build the three-dimensional crystal structures

Copper metal has been used by humans for over 8,000 years. Evidence of the smelting of copper metal dates back to 4500 BC. Copper has high resistance to corrosion, high thermal conductivity and very high electrical conductivity. It has a wide range of applications ranging from the electronics industry to medicine and manufacturing. The construction industry takes advantage of its unique appearance when it becomes oxidized. Copper's lattice structure is face-centred cubic.

▲ Copper is sometimes used as a roofing material. Its characteristic green color is unmistakable. The Statue of Liberty on Liberty Island in New York City is made from hundreds of copper sheets that have changed their form due to the reaction of the copper with air and water over many years. Are there buildings in your country that use copper in their construction?

Iron is an example of a metallic element that has a body-centred cubic crystal structure, and cadmium has a hexagonal close-packed crystal structure. How does the use of models help you to understand the unique patterns that exist in metals at an atomic level?

How is metallic bonding different to ionic and covalent bonding?

Each type of bonding—ionic, covalent and metallic—has its own distinct features. Metallic bonding, like ionic and covalent bonding, can be represented using models which help scientists to explain certain properties which cannot be observed.

The valence electrons involved in ionic bonding are localized on the individual ions. This means that they are not free to move in any of the states of matter. They remain a part of the structure of the ions. In covalent compounds, valence electrons that are the shared pairs of electrons within the bond are again localized. They create electron density in and around the internuclear axis.

So, how do electrons become delocalized within a metal? Metals have low ionization energies—this is the energy required to remove an electron from the outer electron shells. The valence electrons in the outer electron shells of each metal atom are only weakly held to a positively charged nucleus, so they are free to move throughout the metal; this is unlike electrons in ionic and covalent bonding. The properties of metals are therefore different from covalent and ionic compounds.

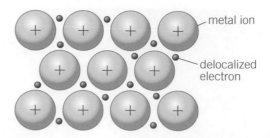

metal ion

delocalized electron

▲ A "sea" of delocalized electrons are free to flow around the positive metal ions which are in structured fixed positions. As the position of the electrons relative to the cations constantly changes, a metallic bond is described as being non-directional. How is this different to the directional bonds present in covalent compounds? Are ionic bonds directional or non-directional?

Models representing metallic bonding help us to explain:

- the structured geometric nature of the positive metal ions
- the delocalized nature of the electrons
- the non-directional nature of the attraction between the positive metal cations and the mobile electrons.

GENERAL

Why are metals malleable and ionic compounds brittle?

We have looked at why ionic compounds are brittle, shattering when a stress is applied to the crystalline lattice structure. The inability of the localized electrons to move within the structure results in like-charged ions repelling each other. Metals are malleable, meaning that they can be shaped by hammering or by a mechanical press without losing structural integrity. As the metal cations are surrounded by electrons, the positive ions can slide over and past one another without causing disruption to the metallic bonding.

▶ Metals are versatile materials that can be hammered into different shapes and drawn into wire. How are these properties different to those of ionic compounds?

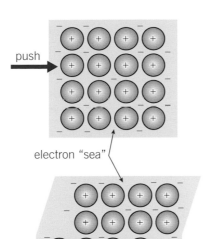

▲ Metallic bonding is not affected when stress is applied to the metal. The positive ions can move over one another without repulsion because the electrons are free to move too

What is an alloy?

GENERAL

An alloy is a macroscopic homogeneous mixture (see Chapter 6, Form) of two or more metals, or a metal combined with what is called an alloying element. The alloying agent is typically made from one or more non-metals. They are classified as mixtures, but combine in such a way that they cannot be separated by mechanical methods. Alloys are important substances that have a wide variety of applications.

Alloy	Composition	Applications
Brass	Copper and zinc	Locks, door handles, musical instruments
Steel	Iron, carbon and another metal such as tungsten	Civil construction such as roads, railways, buildings and bridges
Stainless steel	Iron, nickel and chromium	Building industry, kitchen utensils

The addition of different atoms or small amounts of an alloying agent causes a change in the properties of an alloy compared to the parent metallic element. Alloys can have:

- greater strength

- greater resistance to corrosion

- greater magnetic properties.

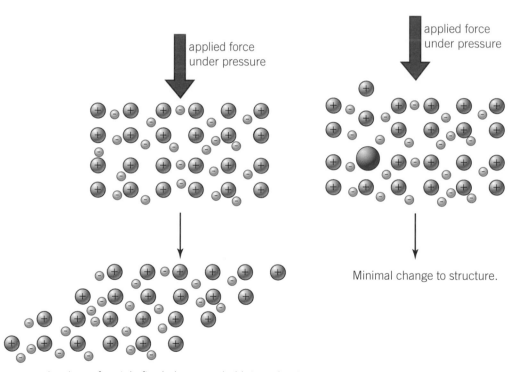

applied force under pressure

applied force under pressure

Minimal change to structure.

▶ The addition of different atoms disrupts the regular lattice structure

structure of metal after being pounded into a sheet

The malleability of metals is the result of their ability to maintain the strength of their non-directional metallic bonds when a force is applied to the metal and the metal cations slide past one another. The disruption caused by the addition of a different atom means that the cations cannot easily slide past one another. The alloy is less malleable and considerably stronger.

Pure aluminium is a soft, corrosion-resistant metal that is both ductile and malleable. It is used in a wide variety of products including soft drinks cans, window frames, lighting fixtures,

car bodies and cooking appliances. It has very high electrical conductivity so it is one of the main components in high voltage power transmission cables. It is often used in alloy materials to improve its strength; for example, lightweight aluminium alloys are used to make aircraft wings and fuselage.

▲ Over the last 40 years there have been significant increases in the number of people flying. To meet this demand, aircraft manufacturers are building greater numbers of aircraft using more resources such as aluminium alloys. Is this a fair and equitable use of the planet's resources if not all people can afford to fly?

Combine knowledge and understanding to create new perspectives

Summarize the key facts in relation to ionic, covalent and metallic bonding in a format of your own choosing. Apply your scientific knowledge and understanding of the concepts covered in this chapter to compare and contrast the formation, and physical and chemical properties. Include information on the following:

- species involved in each type of bonding

- the role of valence electrons

- a description of bond formation

- a discussion of the structures formed

- physical properties such as melting points, boiling points, volatility, solubility in water, and electrical and heat conductivity.

Summative assessment

Statement of inquiry:

Molecular modelling is used for the visualization of chemical structures displaying their orientation in space and time.

Introduction

In this assessment we look at the models that represent the main types of chemical bonding, and how understanding the position and orientation of atoms and ions helps us to explain and predict the chemical and physical properties of ionic, covalent, and metallic substances.

 Chemical bonding and its effect on the properties of materials

1. Ions formed by metals and non-metals are held together by an electrostatic attraction of unlike charges.

   a) Determine the number of valence electrons and the charge formed on each of these elements:

   sodium (group 1)   calcium (group 2) aluminium (group 13)

   nitrogen (group 15) oxygen (group 16) bromine (group 17). [6]

   b) Explain how the metallic and non-metallic ions from these groups can combine to form ionic compounds. [3]

   c) Using the elements given in part a) construct formula and give the name of the ions and the resulting ionic compound from the following combinations:

   i) group 1 and group 17

   ii) group 1 and group 16

   iii) group 2 and group 16

   iv) group 2 and group 15

   v) group 13 and group 16

   vi) group 13 and group 17. [18]

2. Use Lewis symbols and structures to explain the formation of the following covalently bonded compounds. [10]

   a) Methane

   b) Hydrogen bromide

   c) Phosphorus pentachloride

d) Oxygen difluoride

e) Carbon dioxide

The water molecule has two valence electrons involved in covalent bonding with two hydrogen atoms and two lone pairs of electrons or non-bonding electrons.

In this water molecule, there are electrons involved directly in bonding and the lone pairs of electrons not involved in the formation of the covalent bond.

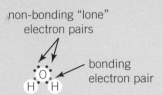

non-bonding "lone" electron pairs

bonding electron pair

3. Consider the compounds in question 2. Can you make the distinction between electrons involved directly in bonding and lone pairs of electrons not required for the formation of the covalent bond? Which molecules in question 2 have lone pairs of electrons and how many? [5]

4. Outline the difference between ionic, covalent and metallic bonding focusing on the following:

   a) electrical conductivity as a solid [3]

   b) melting point [3]

   c) solubility in water. [3]

5. Draw a diagram of metallic bonding and use this to explain the following physical properties of a metal:

   a) electrical conductivity [2]

   b) malleability [3]

   c) ductility (the ability to be drawn into a wire). [2]

Investigating conductivity

The conductivity of an aqueous solution is dependent on the solute and a number of other factors.

You are provided with the following chemicals and apparatus:

- 0.5 mol dm⁻³ potassium chloride
- 1.0 mol dm⁻³ potassium chloride
- 1.5 mol dm⁻³ potassium chloride
- 1 mol dm⁻³ calcium dichloride
- 1 mol dm⁻³ aluminum trichloride
- 1 mol dm⁻³ hydrochloric acid
- 1 mol dm⁻³ ethanoic acid (CH_3COOH, better known as acetic acid)
- Conductivity probe or equivalent
- Standard laboratory glassware
- Safety glasses

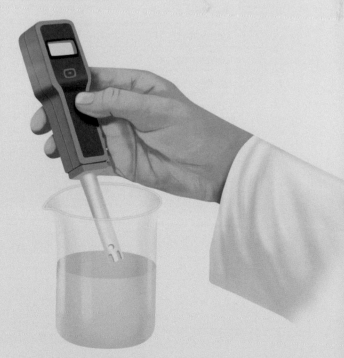

▲ Probes like this can be used to measure the conductivity of solutions

The following hypotheses are all correct. Choose one that you will investigate:

a) An increase in the concentration of a solute will increase the conductivity.

b) The conductivity of a solution is dependent on the number of ions present in a compound.

c) There is a difference in the conductivity of a strong acid when compared to a weak acid.

Support the reason for your choice with scientific reasoning.

6. Design an experiment to test your hypothesis. The method should include:

• the independent and dependent variables and other variables being controlled

• recording quantitative and qualitative observations. [7]

 ## How does chemical bonding relate to electrical conductivity?

Following the design of their own investigation on conductivity, students were asked to perform a different investigation on the conductivity of a wide range of substances. The results of this student's observations have been tabulated below.

Substance	Brightness of lamp	Substance	Brightness of lamp
Distilled water	None	Solid sugar	None
Tap water	Dull light	Graphite	Dull light
Copper metal	Bright light	0.1 M hydrochloric acid	Dull light
Dilute sodium chloride solution	Dull light	Sugar solution (tap water)	Dull light
Concentrated sodium chloride solution	Bright light	Ethanol	None
Solid sodium chloride	None		

The students tested a range of solids and liquids

7. Discuss the differences observed in the level of conductivity in the solids. Your answer should include:

a) comparison of the type of bonding present in each of these solids [3]

b) an explanation of the features of a solid that enables it to conduct an electric current. [2]

8. Examine the results from the testing of distilled and tap water and explain these results using scientific reasoning. [3]

9. Organize the liquids into groups of good, poor and non-conductors.

   a. What features do the good conductors of electricity have in common? [2]

   b. Suggest how the level of conductivity of the poor conductors could be improved. [2]

 ## Graphene – the wonder material

Graphene has endless applications and will allow technologies we may not yet have imagined to be developed. A carbon allotrope, graphene was discovered in 2004 by two British scientists, Sir Andre Geim and Sir Konstantin Novoselov (who went on to win the Nobel Prize for physics for the discovery). Graphene is the material of our future. What qualities make graphene so exciting? Its strength, lightness, conductivity, flexibility, and thinness are some key properties of graphene. Imagine a sheet of graphene the size of an A4 piece of paper that requires the power of many large trucks to tear it apart. Graphene is 300 times stronger than steel and is harder than diamonds.

Graphene-based inks can be printed on clothing; graphene is bendable, light and transparent allowing for use in wearable electronics and medical sensors. Processors for computers presently operate at approximately 2.9 GHz. When using graphene in the processors they operate at 100 GHz. Scientists have developed graphene aerogel, the lightest substance ever made. Batteries using graphene last ten times longer and recharge in a few minutes. LED lighting with graphene filaments is brighter, lasts longer and is cheaper than traditional alternatives. Graphene added to paint can stop the corrosion, or rusting of ships and cars. When graphene is used to make superfine sieves for the process of desalination, one of the world's biggest challenges may be solved – the provision of sufficient fresh, clean water.

Research on the applications of graphene is taking place worldwide, as companies and scientists race to register patents on new products and techniques using graphene. Being made from carbon, which is an abundant element, gives graphene another advantage over other materials. At present, demand for the product exceeds supply, despite the fact that graphene costs approximately $100 per gram to manufacture. Graphene is only one atom thick but will change our lives, now and in the future.

10. List five characteristics, or properties of graphene. [5]

11. Describe three examples of how graphene could be used in your everyday life in the future. [3]

12. The impact of scientific research is far reaching. Explain why the use of graphene in desalination plants could be very significant. [3]

13. Utilizing the characteristic properties of graphene, suggest three present technologies that would benefit from the use of graphene in the production process. Give reasons for your answer. [6]

10 Movement

▲ The Earth is in perpetual motion orbiting the Sun. Since Sputnik 1 was launched into space in 1957, scientific and technological innovations have resulted in our planet being orbited by thousands of artificial satellites. These are used for communication, scientific research, monitoring global weather, military purposes, navigation and astronomy. The International Space Station is an international, collaborative, scientific project that pushes the boundaries of human endeavours and which requires advances in science and technology. NASA uses nickel–hydrogen battery cells to power their satellites. These rechargeable electrochemical cells are more reliable and long-lasting than traditional dry cell batteries. The cells convert chemical energy into electrical energy through the passage of electrons and ions within the system. How else have we benefited from technology developed for space programmes?

UV

The color of the Sun when viewed from the Earth is a consequence of the scattering of the low-wavelength light by the Earth's atmosphere. The Sun emits many different types of electromagnetic radiation, mostly visible light, ultraviolet (UV), and infrared rays. Photons emitted by the surface of the Sun move through the vacuum of space to reach our planet. Excessive exposure to UV rays from the Sun can lead to many degenerative diseases of the eye. Some spectacles will have chemical compounds such as silver chloride embedded within the lens. When exposed to UV light, these compounds are oxidized. This causes the lens to darken and shields our eyes from UV rays. What other chemical reactions are initiated by UV light?

The School of Mathematics and Physics at The University of Queensland, Australia is conducting the world's longest running uninterrupted laboratory experiment. Started in 1927, the experiment is investigating the fluidity and high viscosity of pitch, a derivative of petroleum or coal tar. Its physical properties are unique in that it demonstrates a solid, brittle nature at room temperature, shattering if hit with a hammer. However, it is in fact a liquid with a viscosity calculated to be 230 billion (2.3×10^{11}) times that of water. A drop of pitch has fallen from the funnel just nine times since the start of the experiment. What other substances undergo movement so slow that it is basically imperceptible?

The Earth is in constant movement, spinning on its axis and orbiting the Sun at enormous speeds. It is made up of four distinct layers: the crust, mantle, outer core and inner core. Molten rock or lava formed in the outer core can erupt from the surface of the planet through volcanoes. Fast-moving lava is virtually unstoppable, resulting in the reshaping of the Earth's crust. The Hawaiian Islands experience constant seismic activity, with many of the islands being the product of lava moving from the interior of the planet to the surface, then cooling and solidifying. What other parts of the world experience similar activity?

Key concept: Change

Related concept: Movement

Global context: Scientific and technical innovation

Introduction

Movement is the act, process or result of moving from one location or position to another. Matter is in constant movement relative to its surroundings. Often we are aware of this movement through the observations we make using our senses. Solvation, or dissolving, is the process where individual ions or molecules are surrounded by the solvent molecules, typically water. While we cannot see the individual water molecules interacting with the salt being dissolved, we can observe the product of this movement of particles: the solid particles disappear as they dissolve into the liquid. Gases move from areas of high concentration to low concentration. The diffusion of gases within the air enables us to smell roses or the food we are cooking. In an equilibrium system, we cannot see the interacting particles but we can see the change in the equilibrium position with observations. The process of melting and freezing water is an example of a thermal equilibrium. As we lower the amount of heat in the system, more ice is observed, however, with an increase in the amount of heat, more liquid water is evident. We can observe the change in state, but we cannot directly observe the increase or decrease in vibrational movement of the individual water molecules.

From the kinetic theory of matter we know that atoms, molecules and ions are in constant motion. Studying their movement requires

▼ The grapefruit battery is an example of an electrochemical reaction

Statement of inquiry:

The changes we observe in a chemical system can help us to infer information about the movement of molecules and their properties.

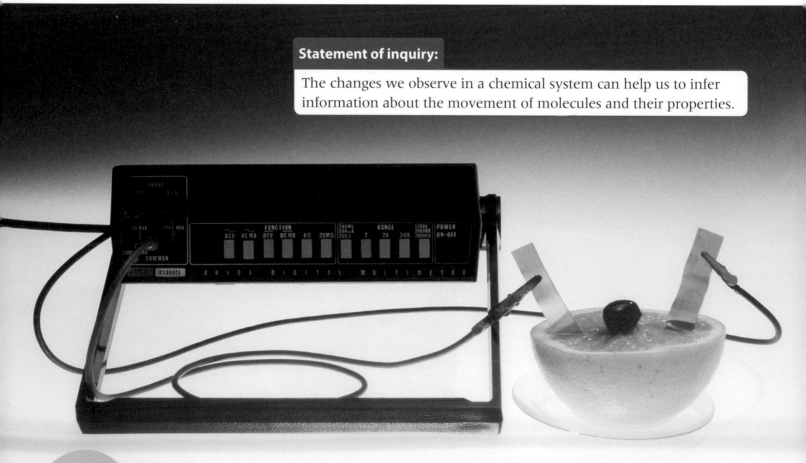

work on a completely different scale – a scale so small that it is difficult to imagine. Chemists use models to help develop and communicate their ideas. When it comes to understanding the movement of electrons between molecules and atoms, the scale is smaller still. Often it is the changes observed that help us to understand the movement. For this reason, the key concept of this chapter is change.

In chemistry, we examine many theories and phenomena that involve movement such as:

- electrons in motion around the nucleus in atomic orbitals

- the transfer of electrons or the sharing of electrons during the formation of a bond

- the movement of reactants and products into and out of reaction systems in industry

- the conversion of reactants to products during chemical reactions

- the movement in the pH of solutions upon reaction with an acid or base

- passage of electrons and ions in an electrochemical cell

- the movement of gases in the atmosphere, their interactions and impact on the environment.

▲ Ocean yacht racing relies heavily on the movement of currents and wind. It also relies on technology accessed on portable devices to provide satellite navigation, weather information and forecasts. The batteries that power the laptops, satellite phones and other systems onboard a racing yacht convert chemical energy into electrical energy. The reliability, portability and lifespan of these batteries has seen the expansion of technological equipment that can be used as handheld or portable devices. Science and technology are partners in the development of products that often improve our quality of life

▲ Zinc metal after it has been left in copper sulfate solution

What movement takes place during a redox reaction?

When a piece of freshly polished zinc metal is placed into a beaker containing a $1\,mol\,dm^{-3}$ solution of copper sulfate, you immediately observe a reaction taking place. The color of the surface of the zinc changes to brown, and the initial blue color of the copper sulfate solution gradually fades and eventually turns colorless. The reaction is complete when the copper ions have come out of solution, to form copper metal in their solid state.

The reaction between zinc and copper(II) sulfate is an example of a redox reaction (see Chapter 8, Interaction). Reduction and oxidation reactions cannot occur separately, as each is dependent on the other. The zinc and copper(II) sulfate reaction can be described by the equation:

$$Zn(s) + CuSO_4(aq) \rightarrow ZnSO_4(aq) + Cu(s)$$
$$\text{(blue solution)} \qquad \text{(red–brown solid)}$$

To better understand what is happening at a molecular level, and examine the movement of ions and electrons within the system, the chemical equation can be simplified by removing the spectator ions—the sulfate ions. The spectator ions are those that play no part in the reaction and remain unchanged within the system. This results in the following equation:

$$Zn(s) + Cu^{2+}(aq) \rightarrow Zn^{2+}(aq) + Cu(s)$$

The electron transfers take place at the surface of the metal. Zinc loses two electrons to form the cation Zn^{2+}—this is the oxidation reaction. The two electrons cannot move through the solution and instead remain on the surface of the zinc. Here they attract the copper ions which move through the solution towards the zinc and change to form elemental copper.

$$Zn(s) \rightarrow Zn^{2+}(aq) + 2e^- \qquad \text{oxidation}$$
$$Cu^{2+}(aq) + 2e^- \rightarrow Cu(s) \qquad \text{reduction}$$

It is important to understand that these reactions do not occur separately in isolation from each other.

● Elemental zinc is responsible for the reduction of copper(II) ions—it is called a reducing agent.

● Copper(II) ions are responsible for the oxidation of elemental zinc—they are called an oxidizing agent.

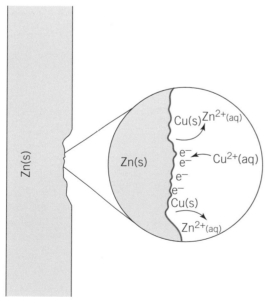

▲ Electron transfer takes place at the surface of the metal

What does the reactivity series tell us?

The reactivity series orders metals according to how easily they are oxidized in their elemental state. If you think of reactive metals, the elements potassium, sodium, and magnesium probably come to mind. Reactivity reflects the ease with which they lose their valence, or outermost, electrons. These electrons move from the metal to the species with which they are reacting. We refer to this as the ease with which they are oxidized.

lithium
potassium
sodium
magnesium
aluminium
manganese
zinc
iron
lead
(hydrogen)
copper
silver
mercury
gold

decreasing reactivity

ease of oxidation increases

◀ The reactivity series ranks elements according to their ease of oxidation

Most reactive metals are found in nature as a compound. For example, many metal ores are found as metal oxides—the metal has been oxidized by elemental oxygen. Other metals are so unreactive there is no movement in the valence electrons. These metals are found in their native state in nature. Precious metals such as gold and platinum are found as pure elements rather than as ores. The metal elements found higher in the reactivity series will displace the positive ions from a compound, forcing them to be reduced. For example, in the reaction between zinc and copper sulfate, zinc is higher in the series and will therefore displace the copper(II) ion from the solution.

1. For each of the following word equations, use the reactivity series to decide if the reaction will proceed spontaneously, as written.

   a) magnesium + hydrochloric acid → magnesium chloride + hydrogen

   b) aluminium + iron(III) oxide → iron + aluminium oxide

   c) magnesium oxide + copper → copper(II) oxide + magnesium

   d) potassium nitrate + zinc → zinc nitrate + potassium

   e) sodium + water → sodium hydroxide + hydrogen

   f) iron(III) oxide + carbon → iron + carbon dioxide

2 Why does gold not react with an acid such as hydrochloric acid?

What is electrochemistry?

Electrochemistry is the field of science that investigates conversions between chemical energy and electrical energy. Movement of electrons and ions is at the centre of these conversions. The increasing global demand for energy has resulted in the development of many different types of electrochemical processes.

A voltaic cell converts chemical energy to electrical energy. The chemical reaction involved in one is commonly a spontaneous, exothermic reaction. Conversely, an electrolytic cell converts electrical energy into chemical energy. The chemical reaction is a non-spontaneous process that requires additional energy from the surroundings to enter the system.

Understanding the redox reactions involved in electrochemical cells has enabled scientists and engineers to manage the changes within them, and as a result develop new applications.

A voltaic cell contains two electrodes: the anode and the cathode. The electrode is a substance through which an electric current enters or leaves the non-metallic part of the electrochemical cell. In both the voltaic and electrolytic cells, oxidation, the process where a species loses one or more electrons, takes place at the anode. Reduction, the process where a species gains one or more electrons, takes place at the cathode. There is a net movement of electrons from the anode to the cathode.

▲ Inert graphite electrodes are used in steel production

What is the role of an electrolyte in an electrochemical cell?

An electrolyte is a substance which produces ions when dissolved in a solvent, resulting in a solution that can conduct charge through the movement of ions in it. Most electrolytes are made from soluble ionic compounds that break up into their ions when dissolved. Chapter 9, Models looks at the conductivity of such compounds. Ionic compounds that fully ionize or dissociate into a solution are called strong electrolytes, while ionic compounds that partially dissociate into ions in solution are weak electrolytes.

In an electrolyte, ions carry the charge between the two electrodes. Generally, electrolytes tend to be ionic substances such as potassium nitrate, sodium chloride and sodium sulfate. Strong and weak acids can also act as electrolytes. Hydrochloric acid, a strong acid, is an effective electrolyte as it completely dissociates into its ions in water. Also defined as a monoprotic acid, one mole of hydrochloric acid will dissociate to produce one mole of hydrogen ions:

$$HCl(aq) \rightarrow H^+(aq) + Cl^-(aq)$$

Sulfuric acid is also a strong acid. However, it is defined as a diprotic acid as one mole of sulfuric acid produces two moles of hydrogen ions when it dissociates in water:

$$H_2SO_4(aq) \rightarrow 2H^+(aq) + SO_4^{2-}(aq)$$

The conductivity of a solution is directly related to the number of ions in solution—the greater the number of ions, the greater the conductivity.

1. Write down some examples of ionic compounds that are soluble in water and dissociate into more than two ions.

Ethanoic acid, a weak acid, only partially dissociates into its ions. It exists in an equilibrium system that favours the undissociated form of ethanoic acid and is therefore not as effective as an electrolyte as hydrochloric acid:

$$CH_3COOH(aq) \rightleftharpoons CH_3COO^-(aq) + H^+(aq)$$

For more about the differences between strong and weak acids see Chapter 3, Consequences.

How can we convert chemical energy into electrical energy?

REDOX

A voltaic cell converts chemical energy into electrical energy, which is then released into the surrounding system to do work. A dry cell battery is one of the most commonly used examples of this.

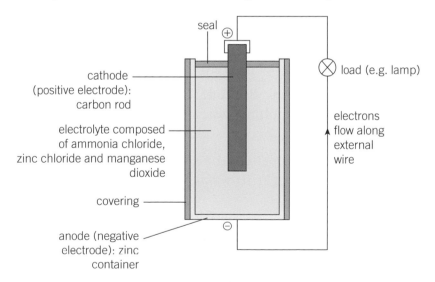

▲ A typical dry cell battery has an electrolyte made of ammonium chloride, NH_4Cl, zinc chloride, $ZnCl_2$ and manganese(IV) oxide, MnO_2. The zinc metal case acts as the anode, and a carbon rod as the cathode. The series of reactions are complex, however, as in all electrochemical cells, oxidation occurs at the anode (in this case the zinc casing), generating electrons that move towards the cathode (in this case the carbon rod) where reduction occurs. A dry cell battery stops producing energy when all of the reactants have been used up. A non-rechargeable battery is also known as a primary cell

Rechargeable battery

A rechargeable battery is known as a secondary cell. It can be recharged by a reversible reaction. The lead-acid battery is used in cars and other types of vehicles. It produces energy capable of starting a combustion engine and is recharged while the motor is running. Its role is primarily to start the engine; it is not required to produce a steady current while the engine is running. The alternator generates energy while the engine is running, powering everything that requires a current within the vehicle and recharging the lead-acid battery.

1. List types of rechargeable batteries that you have encountered and the types of device that they are used in.

2. Evaluate and discuss the advantages and disadvantages of a rechargeable battery system over a non-rechargeable battery system.

3. Suggest why governments encourage individuals to recycle and dispose of batteries responsibly. Describe the type of environmental damage which may occur when batteries are thrown into landfill.

REDOX

Which species are capable of movement in a voltaic cell?

A simple voltaic cell is made of two separate half-cells, with oxidation occurring in one half-cell at the anode and reduction occurring in the other half-cell at the cathode. The cells are separated to allow the movement of electrons through an external circuit from the anode to the cathode so that the electrical energy can be harvested. The circuit in a voltaic cell is completed with the use of a salt bridge which acts as an electrical contact between the two half-cells.

For the simple cell shown in the diagram:

- one half-cell contains a zinc electrode (the anode) placed into an electrolyte solution of $1 \, mol \, dm^{-3}$ zinc sulfate; it contains $Zn(s)$ and $Zn^{2+}(aq)$

- the other half-cell is a copper electrode (cathode) in an electrolyte solution of $1 \, mol \, dm^{-3}$ copper(II) sulfate; it contains $Cu(s)$ and $Cu^{2+}(aq)$.

At this stage, no reaction occurs. The two half-cells are physically separate, and for the oxidation and reduction reactions to proceed, the cells need to be joined to enable the movement of electrons and ions between them. An electrical conducting wire enables electrons to move from the anode to the cathode, and the circuit is completed by the addition of a salt bridge. A salt bridge forms a connection between the two electrolyte solutions and allows the movement of ions between the two cells. Generally, a soluble salt solution is placed within the salt bridge. A nitrate salt is often used.

The overall balanced chemical equation for this reaction can be expressed as an ionic equation, as the anions that form a part of the electrolyte solutions take no part in the reaction—they are spectator ions:

$$Zn(s) + Cu^{2+}(aq) \rightarrow Zn^{2+}(aq) + Cu(s)$$

Oxidation	Reduction
$Zn(s) \rightarrow Zn^{2+}(aq) + 2e^-$	$Cu^{2+}(aq) + 2e^- \rightarrow Cu(s)$

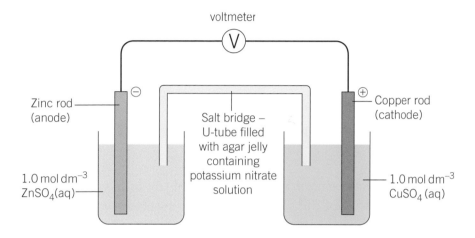

Simple voltaic cell

1 Why do negatively charged anions from the salt bridge move into the anode cell and positively charged cations move into the cathode cell when the half-cells are connected?

  Experiment

Investigating a voltaic cell

⚠️ **Safety**

- Wear safety glasses to protect against any chemical splashes.

- Dispose of the waste solutions appropriately.

Materials

- Two 100 cm³ beakers

- Clean copper, zinc, magnesium and iron electrodes

- 1.0 mol dm⁻³ aqueous solutions of copper(II) sulfate, zinc sulfate, iron nitrate and magnesium sulfate

- Voltmeter

- Electrical wires and connector clips

- Filter paper and 1.0 mol dm⁻³ potassium nitrate solution

Method

1. Half-fill a 100 cm³ beaker with 1 mol dm⁻³ copper(II) sulfate.

2. Half-fill another 100 cm³ beaker with 1.0 mol dm⁻³ magnesium sulfate.

3. Place the copper electrode into the copper sulfate(II) solution and the magnesium electrode into the magnesium sulfate solution.

4. Fold a piece of filter paper into a rectangular strip approximately 1 cm wide. Soak the filter paper in potassium nitrate solution. Place one end into the copper(II) sulfate solution and the other end into the magnesium sulfate solution.

5. Connect a voltmeter between the copper and magnesium electrodes.

6. Record the voltage and your observations in an appropriate table.

7. Construct the overall ionic equation for the reaction taking place and add this to your table.

8. Repeat the method outlined above using other combinations of electrodes and salt solutions.

Questions

1. Draw a labelled diagram of the apparatus and chemicals used in one of the tests you performed. Ensure that you label the polarity of the electrodes, the voltmeter, the anode and cathode, the direction of electron flow in the wires and the salt bridge.

2. Construct half-equations for each of the reactions that occurred at the anode and cathode.

ATL Thinking in context

Utilizing the voltaic cell

Advances in technology are driven by improved understanding of the underlying scientific principles. In turn, improvements in technology help to accelerate scientific knowledge and understanding. Science and technology form an integrated partnership in the modern world. One cannot exist without the other.

Science, technology, engineering and mathematics (STEM) all contribute to economic prosperity and impact our lives through the development of processes and products. Nowadays science is integrated in a much more direct way into the daily lives of a significant proportion of the global population compared to several centuries ago. For example, think how STEM has impacted on education—compare this colorful textbook and other learning aids you use with those used in the 1800s.

1. In a small group, compile a list of voltaic cell applications that impact your daily lives.

2. Discuss how your life would be different if these applications, a product of STEM, did not exist.

How do hydrogen fuel cells work and what is their impact on the environment?

The first simple fuel cell to generate electrical energy from chemical energy was invented in 1838 by Welsh scientist William Grove. His design was later developed into the modern-day phosphoric acid fuel cell. NASA (National Aeronautics and Space Administration) has gone on to develop small-scale, efficient hydrogen fuel cells for use on the International Space Station and satellites, one of many innovations they have developed that has gone on to have more widespread applications.

The hydrogen fuel cell is an electrochemical cell that uses hydrogen and oxygen gases as fuel. The products of the redox reaction are water, electricity and heat:

$$2H_2(g) + O_2(g) \rightarrow 2H_2O(g)$$

reduction reaction at cathode:　　oxidation reaction at anode:
$$O_2(g) + 4H^+ + 4e^- \rightarrow 2H_2O(g) \qquad 2H_2(g) \rightarrow 4H^+ + 4e^-$$

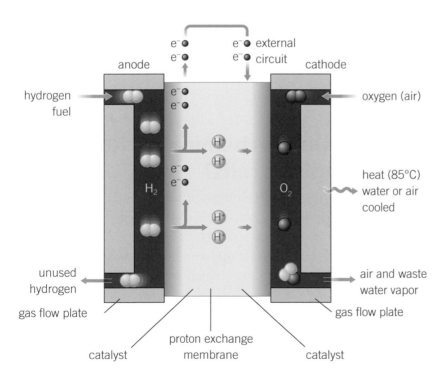

◁ The hydrogen fuel cell generates energy from redox reactions involving hydrogen gas and oxygen gas. Commercially the main source of hydrogen gas is natural gas. How we can produce oxygen gas?

The reactants in a hydrogen fuel cell are renewable and the products are not harmful to the environment. Another major advantage that fuels cells have over conventional batteries is that they continue to convert chemical energy into electrical energy as long as they receive a supply of hydrogen and oxygen.

Advances in the design of these energy cells is focusing on how to supply the raw ingredients that produce the reactants for this electrochemical cell. Molecular hydrogen, H_2, is not that abundant on Earth and needs to be extracted from other compounds such as natural gas and methanol.

Information literacy skills

Collect, organize and present information

You can demonstrate your information literacy skills by finding, analyzing and presenting information in the following ways:

- Make connections between different sources of information.

- Evaluate and select information sources based on how appropriate they are for a specific task.

- Create references.

Electrochemical cells convert chemical and electrical energy through a series of redox reactions. Chemists utilize advances in technology to create innovations and products which impact on our daily lives.

With these skills in mind, consider the following topics:

- Efficiency: In terms of energy production, outline how a hydrogen fuel cell is different to a non-rechargeable dry cell battery.

- Miniaturization: Alkaline batteries are primary, non-rechargeable voltaic cells. State some applications of these small batteries in daily life.

- Fuel sources: Explain the role a hydrogen fuel reformer plays in hydrogen fuel cells.

- Portability: Discuss how portability is an advantage for power sources.

REDOX

How can the electrochemical series help us to construct voltaic cells?

When constructing a voltaic cell, choosing the composition of each half-cell is important in order to achieve a spontaneous reaction that will convert chemical energy to electrical energy. You can use the electrochemical series to determine which of the two selected half-cells will undergo oxidation and which will undergo reduction, relative to each other.

When examining the electrochemical series, you will notice that very reactive metals are found at the top of the series. Metals such as lithium, sodium, potassium, and magnesium all readily lose one or more of their valence electrons to form positive metal ions—they readily oxidize. For this reason, metals high in the series are strong reducing agents due to this tendency to donate electrons.

At the bottom of the electrochemical series is fluorine. It is a very reactive non-metal with a strong tendency to gain an electron and fill its outer electron shell. Because of this strong tendency to gain electrons and become reduced, non-metals high in group 17 are powerful oxidizing agents.

The electrochemical series can be used to predict the behavior of a cell by considering each half-cell.

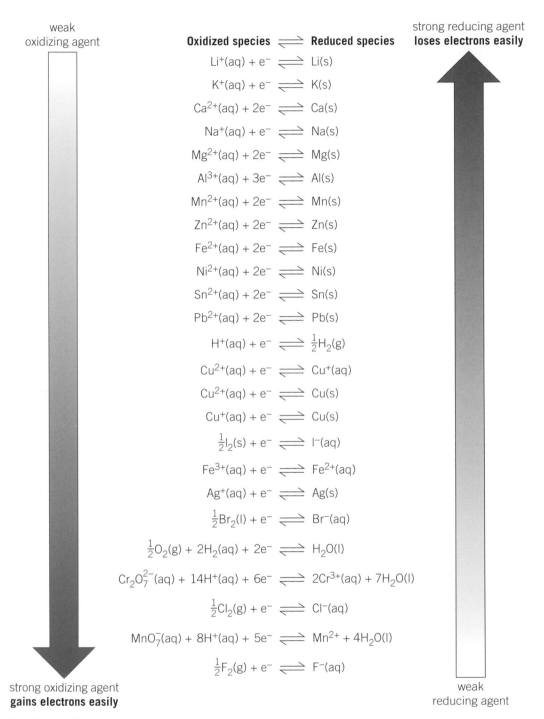

weak
oxidizing agent

Oxidized species ⇌ Reduced species

$Li^+(aq) + e^-$ ⇌	$Li(s)$
$K^+(aq) + e^-$ ⇌	$K(s)$
$Ca^{2+}(aq) + 2e^-$ ⇌	$Ca(s)$
$Na^+(aq) + e^-$ ⇌	$Na(s)$
$Mg^{2+}(aq) + 2e^-$ ⇌	$Mg(s)$
$Al^{3+}(aq) + 3e^-$ ⇌	$Al(s)$
$Mn^{2+}(aq) + 2e^-$ ⇌	$Mn(s)$
$Zn^{2+}(aq) + 2e^-$ ⇌	$Zn(s)$
$Fe^{2+}(aq) + 2e^-$ ⇌	$Fe(s)$
$Ni^{2+}(aq) + 2e^-$ ⇌	$Ni(s)$
$Sn^{2+}(aq) + 2e^-$ ⇌	$Sn(s)$
$Pb^{2+}(aq) + 2e^-$ ⇌	$Pb(s)$
$H^+(aq) + e^-$ ⇌	$\frac{1}{2}H_2(g)$
$Cu^{2+}(aq) + e^-$ ⇌	$Cu^+(aq)$
$Cu^{2+}(aq) + e^-$ ⇌	$Cu(s)$
$Cu^+(aq) + e^-$ ⇌	$Cu(s)$
$\frac{1}{2}I_2(s) + e^-$ ⇌	$I^-(aq)$
$Fe^{3+}(aq) + e^-$ ⇌	$Fe^{2+}(aq)$
$Ag^+(aq) + e^-$ ⇌	$Ag(s)$
$\frac{1}{2}Br_2(l) + e^-$ ⇌	$Br^-(aq)$
$\frac{1}{2}O_2(g) + 2H_2(aq) + 2e^-$ ⇌	$H_2O(l)$
$Cr_2O_7^{2-}(aq) + 14H^+(aq) + 6e^-$ ⇌	$2Cr^{3+}(aq) + 7H_2O(l)$
$\frac{1}{2}Cl_2(g) + e^-$ ⇌	$Cl^-(aq)$
$MnO_7^-(aq) + 8H^+(aq) + 5e^-$ ⇌	$Mn^{2+} + 4H_2O(l)$
$\frac{1}{2}F_2(g) + e^-$ ⇌	$F^-(aq)$

strong reducing agent
loses electrons easily

strong oxidizing agent
gains electrons easily

weak
reducing agent

▲ Electrochemical series

Worked example: Voltaic cells

Question

Consider a copper half-cell and an aluminium half-cell, each with the metal in a solution of its own ions (Al^{3+} and Cu^{2+}). Work out a balanced equation for the cell and determine which metal is the anode and which is the cathode.

Answer

First, we identify the possible reactions in each half cell.

	Aluminium half-cell	Copper half-cell
Contains:	$Al^{3+}(aq)$ and $Al(s)$	$Cu^{2+}(aq)$ and $Cu(s)$
Possible reactions:	$Al^{3+}(aq) + 3e^- \rightarrow Al(s)$ or $Al(s) \rightarrow Al^{3+}(aq) + 3e^-$	$Cu^{2+}(aq) + 2e^- \rightarrow Cu(s)$ or $Cu(s) \rightarrow Cu^{2+}(aq) + 2e^-$

Then we can use the electrochemical series to determine which reaction takes place.

The higher a metal is in the electrochemical series, the higher its reducing power and tendency to lose electrons. By looking at the relative position of each reaction in the electrochemical series, you can determine which metal is more reducing and therefore which will give up its valence electrons.

Relative positions:	Aluminium is higher than copper.	
	Aluminium has a stronger tendency than copper to lose its valence electrons.	Copper has a stronger tendency to gain electrons than aluminium.
Half-cell reaction:	~~$Al^{3+}(aq) + 3e^- \rightarrow Al(s)$~~ ✗ $Al(s) \rightarrow Al^{3+}(aq) + 3e^-$ ✓	$Cu^{2+}(aq) + 2e^- \rightarrow Cu(s)$ ✓ ~~$Cu(s) \rightarrow Cu^{2+}(aq) + 2e^-$~~ ✗

So the two half-cell reactions are:

$$Al(s) \rightarrow Al^{3+}(aq) + 3e^-$$
$$Cu^{2+}(aq) + 2e^- \rightarrow Cu(s)$$

We can then work out the overall reaction that takes place in the cell.

As the number of electrons produced and consumed by the two half-cell reactions above is unequal, determine the lowest common factor. In this case it is six (2×3). Therefore, multiply the first equation by two and the second equation by three:

$$\times 2: \quad 2Al(s) \rightarrow 2Al^{3+}(aq) + 6e^-$$
$$\times 3: \quad 3Cu^{2+}(aq) + 6e^- \rightarrow 3Cu(s)$$

The overall equation for this voltaic cell is:

$$2Al(s) + 3Cu^{2+}(aq) \rightarrow 2Al^{3+}(aq) + 3Cu(s)$$

Finally, we need to determine which is the anode and which is the cathode.

To find the anode and cathode remember that the electrode where oxidation occurs is the anode and the electrode where reduction occurs is the cathode.

The aluminium electrode is oxidized (it loses electrons) therefore it is the anode.

The copper ions, Cu^{2+}, in the copper half-cell are reduced (gain electrons), and therefore the copper electrode is the cathode.

1. For voltaic cells made up of the following half-cells, use the electrochemical series to work out the overall balanced equation for the cell.

   a) Half-cell 1: $Zn(s)$ and $Zn^{2+}(aq)$, half-cell 2: $Pb(s)$ and $Pb^{2+}(aq)$

   b) Half-cell 1: $Cu(s)$ and $Cu^{+}(aq)$, half-cell 2: $Fe(s)$ and $Fe^{2+}(aq)$

   c) Half-cell 1: $Mg(s)$ and $Mg^{2+}(aq)$, half-cell 2: $Sn(s)$ and $Sn^{2+}(aq)$

   d) Half-cell 1: $Ni(s)$ and $Ni^{2+}(aq)$, half-cell 2: $Br_2(l)$ and $Br^{-}(aq)$

   e) Half-cell 1: $Na^{+}(aq)$ and $Na(s)$, half-cell 2: $Ag(s)$ and $Ag^{+}(aq)$

Why do we need to protect the surface of metals from air?

REDOX

Oxygen is an essential component for most living organisms. It is also a strong oxidizing agent; oxidation occurs when a substance reacts with molecular oxygen. Most reactive metals exist as metal oxides and are found in ore bodies in this form. How can we use this oxidation reaction to examine the composition of air?

 Experiment

What is the percentage of oxygen in atmospheric air?

 Safety

- Wear gloves as steel wool can result in steel splinters entering the skin.
- Wear safety glasses.

Materials

- Steel wool
- $400\,cm^3$ beaker
- $50\,cm^3$ measuring cylinder
- Distilled water
- Retort stand and clamp
- Protective gloves

Method

1. Wedge a small, loosely packed ball of dampened steel wool into the bottom of a measuring cylinder. Ensure that it remains in place when the cylinder is inverted.

2. Record the appearance of the steel wool.

3. Place approximately $200\,cm^3$ of distilled water into the beaker.

4. Invert the measuring cylinder and stand it in the beaker.

5. Secure the measuring cylinder using a retort stand and clamp.

6. Record the height of the water inside the measuring cylinder using the graduations on the cylinder.

7. Leave the experiment to sit for at least one week, then measure the height of the water in the measuring cylinder again and record the appearance of the steel wool.

Questions

1. Describe the changes you observed in the steel wool.

2. Calculate the percentage change in the amount of air in the measuring cylinder.

3. Explain how the percentage change in the level of air in the measuring cylinder compares with the percentage of oxygen in atmospheric air.

4. Write a balanced chemical equation to describe the reaction that has taken place between the element iron and oxygen.

5. Identify the type of reaction that iron has undergone.

REDOX

How can we convert electrical energy into chemical energy?

An electrolytic cell uses an external power supply to drive a non-spontaneous chemical reaction. The process converts the supplied electrical energy into chemical energy. It differs from the voltaic cells we looked at earlier in that the reaction takes place in a single cell with the anode and cathode positioned beside each other in the electrolyte. The external power supply is basically an electron pump that provides electrons for the redox reaction.

An electrolytic cell also differs from a voltaic cell in that the negatively charged electrode is described as the cathode and the positively charged electrode is the anode. Remember, the anode and the cathode are defined by where oxidation and reduction take place in the cell. A useful mnemonic to remember this is:

AnOx Red**Cat**

An electrolytic cell can be used to electrolyse water and separate it into its constituent elements: hydrogen and oxygen. Pure water does not conduct electricity, so a small amount of dilute hydrochloric acid is added to create an electrolyte.

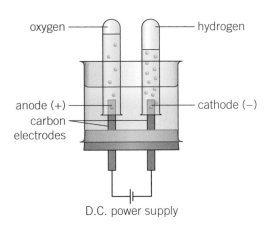

oxygen ——— hydrogen

anode (+) ——— cathode (−)
carbon
electrodes

D.C. power supply

◀ The electrolysis of water produces hydrogen gas at the cathode and oxygen gas at the anode

At the cathode, hydrogen ions present in the solution are reduced to hydrogen gas. Oxygen gas is produced at the anode from the oxidation of hydroxide ions:

$$\text{cathode: } 4H_2O(l) + 4e^- \rightarrow 2H_2(g) + 4OH^-(aq)$$

$$\text{anode: } 4OH^-(aq) \rightarrow O_2(g) + 2H_2O(l) + 4e^-$$

This gives the overall balanced equation:

$$2H_2O(l) \rightarrow 2H_2(g) + O_2(g)$$

This reaction is a simple example of an applied electric current being used to drive non-spontaneous reactions to completion.

Hofmann's voltameter

The Hofmann voltameter was invented in 1866 to separate water into its constituent elements. Three glass tubes are connected by a bridge near their bases. The middle tube is used to fill the apparatus with water and a small amount of electrolyte (for example, dilute sulfuric acid). Each of the other tubes has a platinum electrode sealed into its base and stopcocks at the top.

Consider the balanced equations for the redox reactions occurring at the cathode and the anode in a Hofmann voltameter.

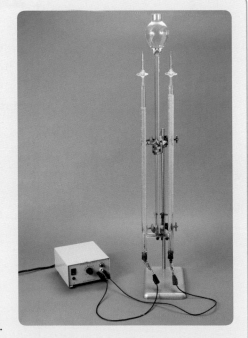

1. Describe how the pH of the solution surrounding the anode and the cathode will change as the reaction proceeds. If a few drops of universal indicator solution are added to the electrolyte, determine what color changes would be observed at each electrode. Explain your answer.

2. Identify which column contains the anode and which the cathode. Explain why there is twice as much gas in one column compared to the other.

How does a molten ionic salt conduct an electric current?

So far we have looked at electrochemical cells with an electrolyte that is an aqueous solution of a metal salt and active metal electrodes, but another possibility is to use a molten solution of a metal salt with inert electrodes.

A solid ionic salt does not conduct an electric current as the ions are locked in place within the crystalline lattice structure. Electrons that move from the metal to the non-metal, creating positively and negatively charged ions, are localized within the ions and cannot move (see Chapter 9, Models). When an aqueous solution of the salt is created, or a solid ionic compound is heated to form a molten solution, the ions are free to move and carry the charge between the electrodes.

In the case of a molten salt solution, inert electrodes are placed into the powdered solid ionic salt and heated strongly. The high level of heat will result in a phase change of the solid ionic salt to give a molten salt solution. Metal and non-metal ions are now free to move and carry charge about the solution.

  Demonstration

Electrolysis of molten zinc chloride

⚠️ **Safety**

- This demonstration should be performed in a fume hood.

- Wear safety glasses.

- Follow correct disposal procedures.

Materials

- Large porcelain crucible

- 15 g of zinc chloride

- Tripod stand and clay-pipe triangle

- Bunsen burner

- Retort stand and clamp

- 2 carbon electrodes

- Connecting wires and crocodile clips

- 1 light bulb

- DC power supply (6 V)

Method

1. Set up the apparatus as shown in the diagram.

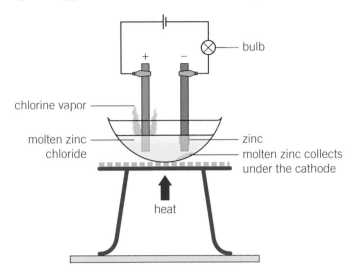

2. Light the Bunsen burner, open the gas sleeve to produce a blue roaring flame and heat continuously until a molten solution of zinc chloride is produced.

Questions

1. What did you observe about the light bulb as the zinc chloride solid began to melt and become molten? Support your answer with scientific reasoning.

2. Construct balanced half-equations for the reactions occurring at the anode and the cathode, and the overall chemical equation.

How are sodium metal and chlorine gas produced commercially?

REDOX

The global demand for chlorine gas continues to rise, as it is the base reagent for a wide variety of commercial products. Chlorine is used as a disinfectant to kill bacteria, and in the production of paint, paper, plastics, insecticides, water pipes, car interiors and electrical wiring insulation. Another major use is in synthetic reactions in the pharmaceutical industry.

Sodium metal is a strong reducing agent used in the production of alloys, and the extraction and purification of more reactive metals such as potassium.

1. In terms of a change of state, what process is involved in the formation of a molten salt?

Sodium chloride is commonly referred to as salt. However, in chemical terms, a salt is defined as a soluble or insoluble combination of a metal ion and a non-metal ion. Sodium

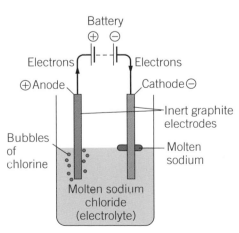

The electrolysis of molten sodium chloride

chloride is extracted in large quantities from seawater by the simple separation process of evaporation. The commercial electrolysis of molten sodium chloride produces approximately 100,000 tonnes of sodium metal annually.

In the molten solution of sodium chloride, the only ions present are sodium ions and chloride ions. When a current is applied to the molten solution, the negatively charged chloride anions move towards the positively charged anode and are oxidized. The positively charged sodium cations move towards the negatively charged cathode and are reduced. Sodium metal is highly reactive and readily undergoes oxidation to form the sodium ion. During the process of electrolysis, a strong electric current is required to force the sodium ion to accept an electron, being reduced in the process:

$$\text{anode:} \quad 2Cl^-(l) \rightarrow Cl_2(g) + 2e^-$$
$$\text{cathode:} \quad 2Na^+(l) + 2e^- \rightarrow 2Na(l)$$

As the oxidation reaction produces two electrons, the reduction reaction at the cathode must be multiplied by 2, to obtain an overall balanced equation:

$$2Na^+(l) + 2Cl^-(l) \rightarrow 2Na(l) + Cl_2(g)$$

1. What observations would you make during the electrolysis of molten sodium chloride?

REDOX

How can electrolysis be used to purify a metal?

The initial extraction of copper from its metal ore produces copper containing sulfur impurities, and electrolysis is used to remove these impurities. The impure copper is cast as an anode electrode ready for the purification process.

The purification process involves electrolysis using impure copper as the anode and a pure copper cathode.

When an electric current is supplied, the impure copper at the anode undergoes oxidation, resulting in the release of copper(II) ions into

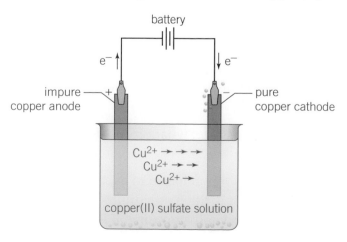

▶ The electrolysis of impure copper to yield pure copper

solution. The copper(II) ions released from the anode, along with the copper(II) ions which are a part of the electrolyte, move towards the cathode where they undergo a reduction process resulting in a layer of pure copper coating the cathode electrode:

anode: $Cu(s) \rightarrow Cu^{2+}(aq) + 2e^-$

cathode: $Cu^{2+}(aq) + 2e^- \rightarrow Cu(s)$

Observations made during this process include the anode losing mass and slowly disappearing while the cathode gains mass and increases in size. Often impurities released from the impure copper anode collect at the bottom of the electrochemical cell.

ATL Thinking in context

Electroplating

One of the most common applications of the electrochemical process of electrolysis is electroplating. Electroplating is binding a thin layer of a metal onto a metal object acting as the cathode electrode. This layer is extremely thin, typically 0.0001 mm in thickness.

1. List some of the applications of the process of electroplating.

The process of electrolysis, and electroplating is dependent on three factors:

- size of the current (in amperes or A)—this is directly proportional to the number of electrons being supplied

- the duration of the process (in seconds)—this is directly proportional to the number of electrons supplied

- the charge on the ion which is being reduced—the higher the charge, the greater the number of moles of electrons required.

2. A current of 10 A is applied for 360 s to a solution of copper(II) sulfate. The same reaction conditions were then applied to a solution of aluminium nitrate. Predict which conditions will result in a greater mass of metal being deposited at the cathode.

3. In the first trial, a current of 5 A was applied for 100 s to a solution of tin(IV) nitrate. In the second trial, the current was doubled to 10 A and applied for 50 s. Evaluate the differences in the reaction conditions.

Where would we be without aluminium?

REDOX

Aluminium is an extremely versatile metal. Properties that make it so useful are that it is: strong and light, an excellent conductor of heat and electricity, easy to form into shapes and structures, an excellent reflector, easy to recycle and resists corrosion. As a result, it is widely used. Aluminium, found as bauxite, or aluminium ore, is one of the most abundant elements in the Earth's crust.

Bauxite is processed using the Bayer process, a series of different chemical reactions, resulting in the production of aluminium oxide (often called alumina). Pure alumina is a white powder with the formula Al_2O_3.

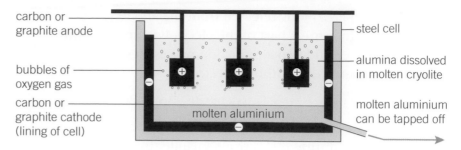

▲ Extraction of aluminium from alumina by electrolysis

Alumina refineries then use the electrochemical process of electrolysis to convert the aluminium oxide into pure aluminium. This process involves a series of redox reactions with aluminium ions being reduced to elemental aluminium.

Alumina or aluminium oxide, Al_2O_3, normally melts at over 2,000°C. The costs to operate an industrial process at this temperature are too high. The addition of cryolite to the alumina creates a mixture which melts at a much lower temperature of approximately 950°C. (Cryolite is a mineral with the systematic name sodium hexafluoroaluminate, Na_3AlF_6.) Molten aluminium oxide breaks up into its ions ready for the process of electrolysis:

$$Al_2O_3(l) \rightarrow 2Al^{3+}(l) + 3O^{2-}(l)$$

At the carbon or graphite cathode, the aluminium ion undergoes reduction, resulting in the formation of liquid aluminium metal:

$$Al^{3+}(l) + 3e^- \rightarrow Al(l)$$

At the graphite anode, oxygen ions are oxidized to elemental oxygen:

$$2O^{2-}(g) \rightarrow O_2(g) + 4e^-$$

The formation of oxygen gas at the surface of the carbon or graphite anode, combined with the high temperature of the reaction mixture, results in oxidation of the anode. The anode slowly breaks down so it needs to be replaced on a regular basis.

$$C(s) + O_2(g) \rightarrow CO_2(g)$$

Aluminium produced using this industrial method is on average 99% pure. Contaminants include iron, silicon and aluminium oxide. Further purification is needed to remove these.

ATL Critical thinking skills

Evaluate evidence

Research three applications of aluminium metal. Analyze the reasons for the choice of aluminium in each in terms of aluminium's physical and chemical properties.

What is diffusion?

Diffusion is defined as the spreading or scattering of gaseous or liquid materials. The process of diffusion is a common phenomenon in our everyday lives. For example, when we smell the fragrance of newly blossomed flowers, food cooking or fumes from vehicles we are experiencing the diffusion or movement of gases. If we leave a tea bag in a cup of hot water without stirring, the color of the water slowly changes as the tea diffuses throughout the water.

▲ Tea bag in hot water

 Demonstration

The reaction between concentrated hydrochloric acid and ammonia

The process of diffusion can be demonstrated by a simple combination reaction between concentrated ammonia and hydrochloric acid.

 Safety

- Wear safety glasses.

- Reaction must be performed in a fume hood.

- Concentrated hydrochloric acid is a highly corrosive substance and should not be handled by students.

- Ammonia is harmful to the respiratory system when inhaled.

Materials

- 6 mol dm^{-3} hydrochloric acid

- 880 ammonia solution

- Cotton wool balls

- Two 50 cm^3 beakers

- Two pairs of metal tweezers

- Dropping pipettes

- 1 m of glass or plastic tubing approximately 3 cm in diameter

- Two clamps and stands

Method

1. Inside a fume hood, clamp the glass or plastic tubing in place.

2. Place a cotton wool ball in each of the beakers.

3. Position the beakers at either end of the tubing.

4. Pour a small quantity of hydrochloric acid onto the cotton wool ball in one beaker and ammonia onto the other cotton wool ball in the other beaker. Immediately replace the lids on the reagent bottles to minimize the release of gases into the work space.

5. Using tweezers, immediately place the cotton wool balls into either end of the tube.

6. Close the safety sash of the fume hood and observe what happens.

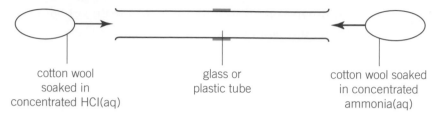

cotton wool soaked in concentrated HCl(aq)

glass or plastic tube

cotton wool soaked in concentrated ammonia(aq)

Questions

1. What is the evidence that a reaction has occurred?

2. The reaction between concentrated hydrochloric acid and ammonia creates a precipitate. Predict the name of the solid formed and its chemical formula.

3. Analyze the results from this experiment and suggest a conclusion about the relationship between the rate of diffusion of a gas and the molecular mass of the gas. Support your statements with scientific reasoning.

The process of diffusion can also involve redox reactions, as shown in the following demonstration.

 Demonstration

Reaction between zinc and iodine and the diffusion of iodine gas in the air

Iodine, a halogen found in group 17, is a purple-black solid at room temperature that sublimates when heat is applied. Sublimation is the change from a solid state directly to a gaseous state (see Chapter 6, Form). If crystals of iodine are placed in a sealed glass vessel, the warmth of your hand will begin the sublimation process and after a short period of time, a purple gas is visible.

 Safety

- Wear safety glasses.

- Perform the reaction in a fume hood or, preferably, outdoors.

- Wear gloves when handling iodine.

Materials

- Iodine solid

- "Gritty" zinc powder/granules

- Distilled water and a pasteur pipette

- Two test tubes or boiling tubes (must be clean and dry)
- Electronic balance

Method

1. Using an electronic balance weigh out 1 g of zinc and transfer it into the test tube.

2. Then weigh out 1 g of iodine and place it in another test tube. Do not mix the solids until you are ready for the reaction to commence.

3. When you have placed the test tubes into the fume hood or taken them outdoors, combine the two solids by placing them onto a watch glass within a metal tin plate. Lightly mix (do **not** grind) with a glass stirring rod.

4. When ready, add a few drops of water using the pasteur pipette and stand well back.

Questions

1. Describe the appearance of the iodine and zinc before the reaction.

2. The reaction between zinc and iodine creates a soluble salt. Predict the name of the salt and its chemical formula.

3. The reaction is strongly exothermic. Explain the role heat plays in producing the formation of the purple iodine gas. Support your statement with scientific reasoning.

◀ When crystals of iodine are gently warmed, iodine vapor is produced

Summative assessment

Statement of inquiry:

The changes we observe in a chemical system can help us to infer information about the movement of molecules and their properties.

Introduction

In this assessment you will examine reactions that convert chemical energy to electrical energy, and electrical energy to chemical energy. The focus of this assessment is the movement of ions and electrons within a chemical system and the factors that influence this. You will need to use your knowledge of electrochemical cells. Finally, you will reflect on developments in knowledge, understanding, and practice in chemistry.

 Electrochemical cells

The following reduction half-reactions have been taken from the electrochemical series. Reactions which are found high in the series have a tendency to undergo oxidation and the reverse equation should be used. Reactions low in the electrochemical series tend to undergo reduction and the forward reaction will occur.

$Zn^{2+} + 2e^- \rightleftharpoons Zn$	Higher in the electrochemical series
$Fe^{2+} + 2e^- \rightleftharpoons Fe$	
$Cu^{2+} + 2e^- \rightleftharpoons Cu$	
$Ag^+ + e^- \rightleftharpoons Ag$	Lower in the electrochemical series

1. **a)** Identify five possible combinations of these half-equations. [5]

    b) Write the complete chemical equation to represent each reaction that you have identified. [10]

2. The mnemonic OIL RIG stands for: oxidation is loss, reduction is gain. This refers to the movement of electrons to and from reacting species during redox reactions. Analyze the equations you constructed in question 1 and state which species in each equation is being oxidized and which is being reduced. [5]

3. Select one of the equations identified above, then follow the steps below.

i) Draw a labelled diagram of a voltaic cell that will generate an electric current. [5]

ii) Label the cathode and the anode. [2]

iii) Indicate all the ions present in each half-cell. [4]

iv) A common choice of electrolyte for a salt bridge is sodium nitrate. Explain why this is a good choice of electrolyte. [2]

v) Indicate the direction of the movement of electrons. [1]

4. Use the electrochemical series to determine the species produced at the anode and cathode during the electrolysis of the ionic salt tin(II) iodide. [2]

 ## Identifying a reactivity series

Your task is to design a series of small experiments to compare the reactivity of different metals during redox reactions. You will be provided with the following:

- freshly cleaned strips of each of the following metals—copper, lead, zinc, and magnesium

- 0.5 mol dm⁻³ solutions of copper(II) nitrate, zinc(II) nitrate, lead(II) nitrate and magnesium sulfate

- test tubes.

5. Develop a hypothesis based on your knowledge of the electrochemical series. [2]

6. Design a scientific experiment to test your hypothesis. The method should include:

- the independent and dependent variables, and other variables being controlled

- a way to record quantitative and qualitative observations. [8]

 ## Identifying a reactivity series – analysis and evaluation

Following a series of redox reactions between a metal and an aqueous salt solution, a student recorded the following color changes to the metal.

	$Cu(NO_3)_2$(aq)	$Pb(NO_3)_2$(aq)	$Zn(NO_3)_2$(aq)	$MgSO_4$(aq)
Cu		no change	no change	no change
Pb	grey to red–brown		no change	no change
Zn	silver-grey to red–brown	grey to black		no change
Mg	silver-grey to red–brown	silver-grey to black	silver-grey to dull grey	

7. Interpret the data and explain the results using scientific reasoning. [6]

8. Evaluate the validity of your method and your hypothesis. [4]

9. Suggest improvements or extensions to your experiment. [3]

Chemistry throughout the centuries

Examining the development of science and technology over the centuries gives us insight into their impact on communities, the environment, and global resources.

▷ Joseph Wright of Derby was an influential painter of the late 18th century. This painting entitled *The Alchemist: in Search of the Philosopher's Stone* depicts a moment in time that forms a part of the history of chemistry

▷ This painting of Louis Pasteur in 1885 by Albert Edelfelt is in the Musée d'Orsay in Paris, France. It depicts Pasteur, a French biologist, microbiologist and chemist at work in his laboratory

10. Discuss how these images portray the journey through time from
alchemy to modern scientific research in the field of chemistry. [6]

11 Patterns

This photo of the Dragon highway intersection in Shanghai, China is typical of the highly complex infrastructure seen in megacities across the world. They are developed to service growing populations and demand for improved lifestyle. Society aims to stay ahead of the demand in providing the necessary infrastructure. How can analyzing the patterns in human migration and movement of traffic guide city development plans for future needs?

Exploration of the planet continues. In a 2014–2015 study of the Amazon, researchers found a new plant or animal species every two days. The 381 new species included 216 plants, 93 fish, 32 amphibians, 20 mammals, 19 reptiles and one bird. It is estimated that 80% of the species that live on our planet are still undiscovered. What patterns exist between the sustainability of wildlife in environments like the Amazon, and the human mining of its resources?

The science of aesthetics is concerned with personal and cultural perceptions of art and beauty. Architectural design and the patterns evident in iconic structures are the result of aesthetic choices which are subject to personal taste. The patterns created by the shell-like roof structures at the iconic Sydney Opera House are considered a masterpiece by many but are less favourably regarded by others. Can you think of any other famous architectural patterns?

The standard model for communicating chemical patterns is two-dimensional diagrams of interacting atoms. Chemical patterns enable us to visualize the features of chemical structures and subsequently analyze the physical and chemical properties of the chemicals. Computation can produce models of these molecular structures in three dimensions, such as insulin, pictured here. Why is it advantageous for chemists to see three-dimensional representations of interacting atoms over two-dimensional diagrams?

Transition metal complexes vary in color as a result of differences in the identity of the metal, the charge on the metal ion and the type and number of species (or ligands) bonded to the metal ion. Patterns in the colors of transition metal complexes have helped us to understand their properties

Introduction

A pattern is a distribution of variables in time and space. This distribution of variables is sometimes identifiable as a unique pattern. Scientists look for patterns to help them make predictions. For example, pharmaceutical drugs with similar structures and functional groups may have a similar effect on the body. Patterns may be obvious or it may require vast amounts of data and computing power to establish links and relationships.

An important feature in chemistry is the search for trends, anomalies and discrepancies in the behavior of elements which can be used to classify them into groups with similar characteristics. We classify elements as metals and non-metals based on what we observe from the way in which they react with other elements and compounds. Identifying these patterns in the properties of elements enables us to make predictions about their behaviour when they are present in new reactions and compounds. For this reason, the related concept of this chapter is patterns.

We sometimes group elements together despite one or more members of the groups not being particularly representative. The d-block elements,

Statement of inquiry:

Chemists look for patterns in the periodic table in order to discover relationships and trends that help them to predict physical and chemical properties.

or transition metals, are found in groups 3–12. The elements in group 12 are anomalies in that they do not display standard transition metal properties. For example, zinc forms white compounds, rather than the characteristic colored compounds of groups 3–11. Understanding the relationship between the properties of a transition metal and the arrangement of electrons in the atom is crucial in making sense of these anomalies. For these reasons, in this chapter we will examine the concept of relationships, and the global context for this chapter is orientation in space and time.

Evidence is of fundamental importance to scientists. Experiments performed in a controlled environment within laboratories enable the scientific community to collect evidence and determine if patterns exist. The use of models and computer-based technology has improved our ability to analyze large amounts of data and establish what patterns exist. The confirmation of the physical and chemical properties of the 118 elements and millions of compounds, through the analysis of experimental or empirical data, allows the accurate prediction of the actions of elements and compounds. This is a great benefit to scientists performing synthesis reactions to produce new chemicals.

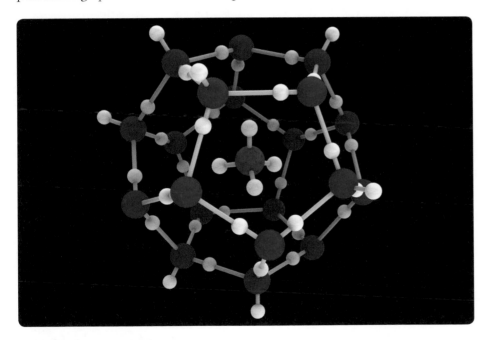

◀ A clathrate is a chemical compound defined by the IUPAC as an inclusion compound in which the guest molecule is in a cage formed by the host molecule or by a lattice of host molecules

What is the arrangement of the modern periodic table?

GENERAL

The periodic table is an invaluable tool for scientists, chemists and students. It contains important information structured in a way that enables us to predict the properties of elements and their reactions, and the formula of many compounds.

There are a few key points to remember about the periodic table.

- The elements of the periodic table are arranged in order of increasing atomic number Z.

▼ The design of the modern periodic table is highly structured. However, the genius of its design is that it is easy to extract information about individual elements and groups of elements. What other examples of tables of information do you use in your daily life?

● Vertical columns, called groups, are numbered from 1 to 18 from left to right. Some of the groups have their own specific names, such as the alkali metals (group 1) and the halogens (group 17).

● The horizontal rows are called periods and are numbered from one to seven, from top to bottom.

Times of discovery

before 1800 | 1900–1949
1800–1849 | 1949–1999
1849–1899

Group	1	2	3	4	5	6	7	8	9	10	11	12	13	14	15	16	17	18
Period																		
1	H 1																	He 2
2	Li 3	Be 4											B 5	C 6	N 7	O 8	F 9	Ne 10
3	Na 11	Mg 12											Al 13	Si 14	P 15	S 16	Cl 17	Ar 18
4	K 19	Ca 20	Sc 21	Ti 22	V 23	Cr 24	Mn 25	Fe 26	Co 27	Ni 28	Cu 29	Zn 30	Ga 31	Ge 32	As 33	Se 34	Br 35	Kr 36
5	Rb 37	Sr 38	Y 39	Zr 40	Nb 41	Mo 42	Tc 43	Ru 44	Rh 45	Pd 46	Ag 47	Cd 48	In 49	Sn 50	Sb 51	Te 52	I 53	Xe 54
6	Cs 55	Ba 56	La* 57	Hf 72	Ta 73	W 74	Re 75	Os 76	Ir 77	Pt 78	Au 79	Hg 80	Tl 81	Pb 82	Bi 83	Po 84	At 85	Rn 86
7	Fr 87	Ra 88	Ac# 89	Rf 104	Db 105	Sg 106	Bh 107	Hs 108	Mt 109	Ds 110	Rg 111							

*58–71 Lanthanides	Ce 58	Pr 59	Nd 60	Pm 61	Sm 62	Eu 63	Gd 64	Tb 65	Dy 66	Ho 67	Er 68	Tm 69	Yb 70	Lu 71
#90–103 Actinides	Th 90	Pa 91	U 92	Np 93	Pu 94	Am 95	Cm 96	Bk 97	Cf 98	Es 99	Fm 100	Md 101	No 102	Lr 103

Just as the biological classification system is designed to group living organisms according to their characteristics, similarly the periodic table groups elements with similar chemical and physical properties into vertical columns.

ATL | **Information literacy and communication skills**

Communicate information and ideas effectively to multiple audiences

Some of the names you encounter in your study of chemistry have unusual origins. For example, the element potassium has the symbol K. Potassium, found in group 1 of the periodic table, is a reactive metal that creates an alkaline solution when it reacts with water. The symbol for potassium originates from the New Latin word kalium, which in turns originates from the Arabic word qali, meaning alkali.

Research the origin of the following group names from the periodic table:

Group 1:   alkali metals

Group 2:   alkaline earth metals

Group 17:   halogens

Group 18:   noble gases

Present your findings in an appropriate manner for one of the following audiences:

- 12–13 year olds who are studying science

- the general public.

1 How does the structure of the modern periodic table enable us to make predictions about the physical and chemical properties of elements based on their relative position within the periodic table?

How can models help us to understand electron arrangements?

GENERAL

Following the discovery of the electron towards the end of the 19th century, Niels Bohr, a Danish physicist, proposed a model to explain how electrons are positioned around the nucleus of an atom. Bohr's model proposed that electrons moved around the nucleus in prescribed orbits, known as orbitals, and that when an electron fell from a higher energy level to a lower energy level, light of a specific wavelength was emitted. The theory that electrons are found in very specific energy levels was confirmed by experimental research. Bohr's model goes on to describe the arrangement and number of electrons in each orbital.

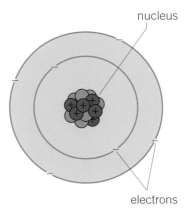

▲ The Bohr theory of the atom is fundamental to our understanding of the electron arrangement within an atom

▲ Bohr's model of the atom explains the frequencies of light seen in the emission spectrum of hydrogen

The energy level of each orbital is described by its principal quantum number n. The first energy level of an atom has the principal quantum number of 1. To determine the maximum number of electrons in each orbital, we use the formula $2n^2$. The first orbital contains a maximum of two electrons, the second level a maximum of eight electrons, and so on. The highest value of n for an element is also the period number in the periodic table.

Principal quantum number n	Maximum number of electrons $2n^2$
1	2
2	8
3	18

◀ Principal quantum numbers of orbitals and the maximum number of electrons

The Bohr model of electron arrangement depicts the electron orbitals as being concentric circles around the nucleus of the atom. This configuration can be used to explain the chemical properties of an element. Understanding the arrangement of the electrons also helps to explain the types of bond that an element can form.

GENERAL

How can we determine the number of valence electrons of an element?

Valence electrons are the electrons found in the outermost orbital of an element. They are very important as they determine an element's chemical properties and bonding behavior. We can determine the number of valence electrons of an element from its group number.

For example, sodium, a reactive metal, is found in group 1 and period 3 of the periodic table. It has an atomic number of 11 and a mass number of 23. This tells you that it has 11 protons and 12 neutrons in its nucleus, and 11 electrons positioned in electron orbitals surrounding the nucleus (see Chapter 7, Function).

Remember: the atomic number is the number of protons in the nucleus of an atom. As an element has no overall charge, it is also equal to the number of electrons. The mass number tells you the total number of protons and neutrons in the nucleus.

▼ The electronic structure of sodium

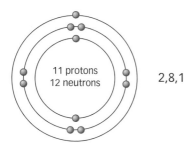

11 protons
12 neutrons

2,8,1

1. How does knowing the number of sub-atomic particles in an atom help us to work out the electron configuration of the element sodium?

Using the Bohr model, the electron arrangement of sodium, Na, is as follows:

- first shell: 2 electrons

- second shell: 8 electrons

- third shell: 1 electron

We use the notation [2,8,1] to describe the electron configuration of sodium. For groups 1 and 2, and Groups 13–18, the final digit of the group number of the periodic table is also the number of valence electrons in elements of that group.

- Groups 1 and 2 have one and two valence electrons, respectively.

- Groups 13–18 have from three to eight valence electrons, respectively (remember to remove the 1 at the front of the group number). For example, silicon in group 14 has four valence electrons. This is evident from its configuration of [2,8,4].

The elements of groups 3–12 are the transition metals. These elements can form ions with different possible charges depending on the species they are reacting with (see Chapter 9, Models). For example, copper can form Cu$^+$ and Cu^{2+} ions. Therefore, we cannot use the earlier "rule of thumb" for working out the valence electrons of transition metals.

Determining the number of valence electrons and electron configuration

Complete the following table to determine the number of valence electrons and electron configuration for the following elements. The first one has been done for you.

Element	Group number	Period number	Electron configuration (using [2,8,8] notation)	Valence electrons
Lithium	1	2	[2,1]	1
Magnesium				
Aluminium				
Carbon				
Nitrogen				
Sulfur				
Fluorine				
Argon				

What is the quantum mechanical model of the atom?

GENERAL

The Bohr model of electron configuration makes good predictions for smaller atoms like hydrogen. However, some of the assumptions that it makes have been found to be incorrect. It has now been replaced by the quantum mechanical theory of the atom, which is a much more complex mathematical model describing the arrangement of electrons in the atom.

> **"**
> Anyone who is not shocked by quantum theory has not understood it.
> **Niels Bohr**
> **"**

◀ If you find the quantum mechanical model challenging, do not panic

The shortcomings of the Bohr model include:

- the idea that an electron in an orbital around the nucleus is like the orbit of a planet around the Sun, in that you can predict its position.

- the assumption that energy levels or orbitals are circular or spherical in nature—in fact they have a wide variety of shapes.

- the idea that electron orbitals exist even when there are no electrons present in a particular energy level (in the way that the levels of a multistory car park exist even when there are no cars there)—we now understand this differently and use the term orbital to describe a region in space where there is a very high probability of finding an electron. If there are no electrons in a particular energy level, then the orbital does not exist.

An electron does not have an exact location; rather there is a region in space where it is highly likely that we can find it. Although we cannot say where the electron is, we can say what energy it has if we know which orbital it is in.

There are four types of orbitals or energy sublevels: *s*, *p*, *d* and *f*.

The *s* orbital is spherical. It can contain up to two electrons. It defines a region in which there is very high probability of finding up to two electrons.

The *p* orbitals are made up of three orbitals that are identical in shape and energy. Each orbital can contain up to two electrons so in total the *p* orbitals can contain up to six electrons.

If you go on to study chemistry at a higher level, you will learn that the five *d* orbitals can contain up to 10 electrons and the seven *f* orbitals up to 14 electrons.

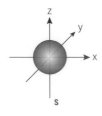

▲ An *s* orbital

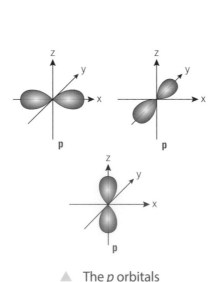

▲ The *p* orbitals

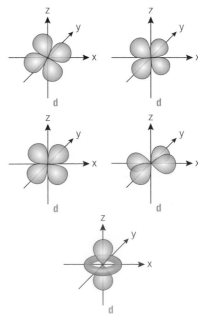

▲ The *d* orbitals

With the exception of the first main energy level, which contains just one s orbital, each main energy level is made up of several orbitals or energy sublevels with similar, but not identical, energies. The table below shows the orbitals for the first few energy levels.

Energy level	Sub levels				Total electrons
1	one s orbital				2
2	one s orbital	three p orbitals			8
3	one s orbital	three p orbitals	five d orbitals		18
4	one s orbital	three p orbitals	five d orbitals	…	

▲ Orbitals making up each of the main energy levels

The following diagram will help you to visualize the energies of the first three main energy levels and the beginning of the fourth level.

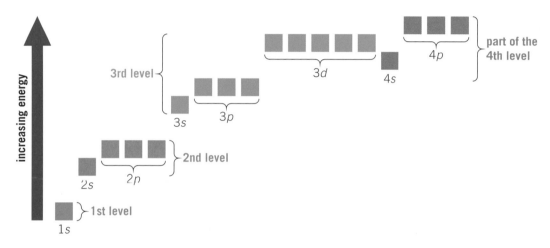

▲ Relative energies of the first three main energy levels and the start of fourth level (not to scale). Remember: each orbital can contain up to two electrons

The atomic number of an element tells you its total number of electrons. So how do the electrons fit into the different orbitals? There are two basic rules to follow when working out the electronic configuration of an element using the quantum mechanical theory:

● Electrons fill the lowest possible energy orbital.

● Any orbital can hold a maximum of two electrons.

In the following examples refer back to the diagram above.

Consider helium, which has atomic number 2 – it has two electrons. The lowest energy level is 1s which can take both electrons. This is written as follows:

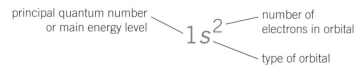

Now consider fluorine, which has atomic number 9, so it has nine electrons. Filling up from the lowest energy level gives:

- 2 electrons in $1s$
- 2 electrons in $2s$
- 5 electrons in $2p$

This is written as $1s^2 2s^2 2p^5$. You can check your electron configuration by adding all the superscript values to make sure that they equal the total number of electrons ($2 + 2 + 5 = 9$).

Now consider calcium, which has atomic number 20, so it has 20 electrons. Filling up from the lowest energy level gives:

- 2 electrons in $1s$
- 2 electrons in $2s$
- 6 electrons in $2p$
- 2 electrons in $3s$
- 6 electrons in $3p$

Now take care! If you look at the diagram on the previous page, the next lowest level after $3p$ is $4s$, not $3d$, and so the remaining two electrons occupy the $4s$ orbital. This is written $1s^2 2s^2 2p^6 3s^2 3p^6 4s^2$.

To form a Ca^{2+} ion, the two outer valence electrons are lost giving a configuration of $1s^2 2s^2 2p^6 3s^2 3p^6$.

The following table gives the configuration for the first ten elements.

Element	Atomic number	Electron configuration
H	1	$1s^1$
He	2	$1s^2$
Li	3	$1s^2 2s^1$
Be	4	$1s^2 2s^2$
B	5	$1s^2 2s^2 2p^1$
C	6	$1s^2 2s^2 2p^2$
N	7	$1s^2 2s^2 2p^3$
O	8	$1s^2 2s^2 2p^4$
F	9	$1s^2 2s^2 2p^5$
Ne	10	$1s^2 2s^2 2p^6$

▲ Full electron configuration for the first ten elements

1. Determine the full electron configuration for the following atoms and ions:

   a) sodium (Z = 11)

   b) aluminium (Z = 13)

   c) chlorine (Z = 17)

   d) vanadium (Z = 23)

   e) copper (Z = 29)

   f) bromine (Z = 35)

   g) oxide ion, O^{2-} (for elemental oxygen, Z = 8)

   h) magnesium ion, Mg^{2+} (for elemental magnesium, Z = 12)

   i) sulfide ion, S^{2-} (for elemental sulfur, Z = 16)

   j) potassium ion, K^+ (for elemental potassium, Z = 19)

It is the outer valence electrons that are responsible for an element's chemical properties, so it can be helpful just to look at the outer shell.

Group

	1	2	3	4	5	6	7	8	9	10	11	12	13	14	15	16	17	18
1	H 1	*s* block																He 2
2	Li 3	Be 4											B 5	C 6	N 7	O 8	F 9	Ne 10
3	Na 11	Mg 12					*d* block						Al 13	Si 14	P 15	S 16	Cl 17	Ar 18
4	K 19	Ca 20	Sc 21	Ti 22	V 23	Cr 24	Mn 25	Fe 26	Co 27	Ni 28	Cu 29	Zn 30	Ga 31	Ge 32	As 33	Se 34	Br 35	Kr 36
5	Rb 37	Sr 38	Y 39	Zr 40	Nb 41	Mo 42	Tc 43	Ru 44	Rh 45	Pd 46	Ag 47	Cd 48	In 49	Sn 50	Sb 51	Te 52	I 53	Xe 54
6	Cs 55	Ba 56	La* 57	Hf 72	Ta 73	W 74	Re 75	Os 76	Ir 77	Pt 78	Au 79	Hg 80	Tl 81	Pb 82	Bi 83	Po 84	At 85	Rn 86
7	Fr 87	Ra 88	Ac# 89	Rf 104	Db 105	Sg 106	Bh 107	Hs 108	Mt 109	Ds 110	Rg 111	Cn 112	Uut 113	Fl 114	Uup 115	Lv 116	Uus 117	Uuo 118

p block

f block

*58–71 Lanthanides	Ce 58	Pr 59	Nd 60	Pm 61	Sm 62	Eu 63	Gd 64	Tb 65	Dy 66	Ho 67	Er 68	Tm 69	Yb 70	Lu 71
#90–103 Actinides	Th 90	Pa 91	U 92	Np 93	Pu 94	Am 95	Cm 96	Bk 97	Cf 98	Es 99	Fm 100	Md 101	No 102	Lr 103

Period

▲ Periodic table with the valence electron orbitals shown. The pattern of these orbitals divides the periodic table into four blocks: the *s* block, the *p* block, the *d* block, and the *f* block

What does the ionization energy of an element represent?

Ionization energy is defined as the minimum energy required to remove an electron from a neutral gaseous atom. It is an indication of the strength of attraction the positively charged nucleus of an atom has for the outermost valence electrons of an atom. The value of the ionization energy is always positive as the atom requires an input of energy to remove an electron – this is an endothermic process. The first ionization energy of an atom X can be expressed by the following chemical equation:

$$X(g) \rightarrow X^+(g) + e^-$$

The metal lithium has an atomic number of three and an electron configuration of [2,1]. It needs to lose one electron to attain a noble gas configuration (see Chapter 9, Models). The equation to describe the first ionization energy of lithium is:

$$Li(g) \rightarrow Li^+(g) + e^-$$
$$[2,1] \quad [2]$$

Comparing the first ionization energies of elements in the same group of the periodic table, or elements across a period, reveals trends in the data, which in turn informs our understanding of the interactions between the nucleus and the valence electrons.

Metals like to lose electrons as their ionization energies tend to be lower than non-metals which do not like to give up their electrons. Non-metals like to gain electrons to complete the outer electron shell and attain a noble gas electron configuration.

1. How can we use the value of the ionization energy of an atom as an indicator of the level of reactivity of an element?

2. What factors determine the relative size of the ionization energy of an element?

What is the pattern of first ionization energy within a group?

What is shielding?

Imagine being part of a large group sitting in concentric circles around a log fire in very cold weather. If you are furthest from the fire, consider how your experience differs from a friend who is sitting closest to it. The radiant energy from the fire is blocked by people sitting in front of you and you will only feel a small amount of the heat energy being generated by the fire.

Similarly, electrons found in the orbitals surrounding an atom experience different levels of attraction towards the positively charged nucleus. Shielding is the term used to describe the effect lower energy, fully filled electron shells have on the level of attraction between the nucleus and the valence electrons.

1 A campfire generates radiant heat that will keep you warm on a cold winter's night. Additional wooden logs will increase the amount of radiant heat produced. Using this analogy, what happens to the nucleus of an atom as the number of protons increases?

Comparing the relative atomic radii of elements gives a good insight into what is happening at an atomic level.

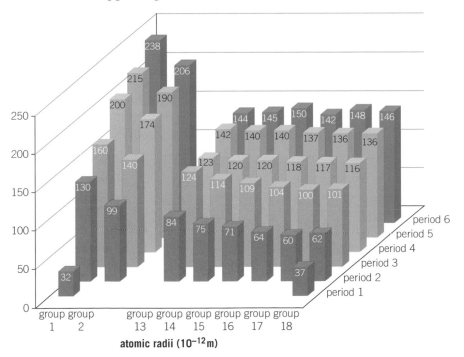

◀ The atomic radius increases down a group and decreases across a period

atomic radii (10^{-12} m)

For example, consider the alkali metals of group 1 in the diagram above. As you move down the group, starting at lithium in period 1, the atomic number increases and the outer electron shell that the valence electrons occupy is at an increasingly higher energy level. Each successive energy level is further away from the nucleus, so the atomic radius of the atom increases. Additionally, the level of shielding between the valence electrons and the nucleus also increases down a group. You can see this if you look at group 1 on the chart of atomic radii.

With increasing levels of shielding as you move down the group, the hold on the valence electrons by the nucleus decreases. This decreasing level of attraction between the nucleus and the valence electrons is evident in the trend of first ionization energy as you move down a group.

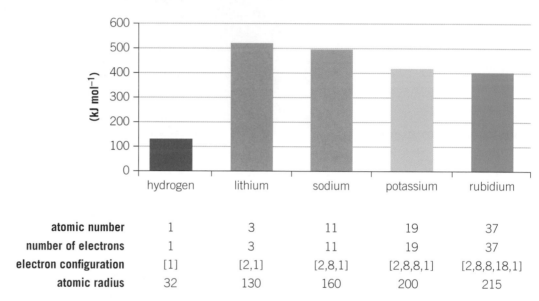

atomic number	1	3	11	19	37
number of electrons	1	3	11	19	37
electron configuration	[1]	[2,1]	[2,8,1]	[2,8,8,1]	[2,8,8,18,1]
atomic radius	32	130	160	200	215

▲ Apart from hydrogen which is a special case, the general trend is a decrease in the first ionization energy down group 1

The ease with which a metal element loses an electron is indicative of the level of reactivity. This means that the decrease in the first ionization energy as you descend group 1 results in an increasing level of reactivity. Put another way, the weaker the hold of the positively charged nucleus on the outermost valence electron, the easier it is for the atom to lose it.

Metals "like" to lose one or more electrons to attain a noble gas electron configuration (see Chapter 8, Interaction). Remember that these electrons are not destroyed—they are transferred to the atoms of elements which are trying to gain an electron, such as non-metals.

What is the pattern of the first ionization energy across a period?

Consider the trend in ionization energy across period 3.

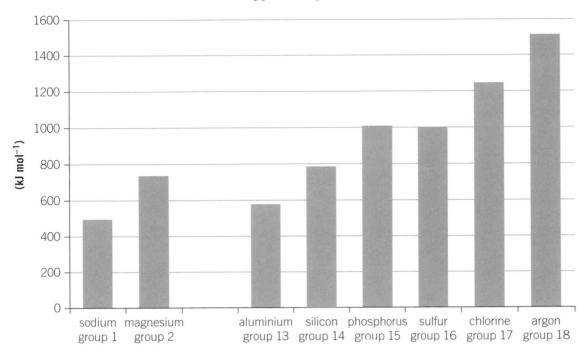

As you move across a period (moving horizontally across a row) in the periodic table, the level of shielding remains constant. For each successive element, the number of electrons in the outermost orbital increases by one. Each additional electron is placed into an orbital with a similar energy level; however, the positivity of the nucleus is increasing for each successive element, as the number of protons in the nucleus increases. With an increasing positivity in the nucleus, and a constant level of shielding, the hold on the outermost valence electrons increases as you move across the period. The positivity of the nucleus is referred to as the effective nuclear charge and it increases from left to right across a period.

A consequence of this increase in attraction between the nucleus and the outermost electrons is that the electrons are harder to remove from the atom, as shown by the general increase in the first ionization energy across the period. The two exceptions to this trend – the decrease in ionization energy at aluminium and sulfur – are explained at DP level. This pattern in ionization energy from metals on the left-hand side to non-metals on the right-hand side of the periodic table helps to explain why metals tend to lose electrons and form cations, while non-metals have a tendency to gain electrons and form anions. The formation of positively charged metal ions and negatively charged non-metallic ions leads to the formation of compounds held together by ionic bonding (see Chapter 9, Models).

The halogens (group 17) are non-metals that have a wide variety of physical and chemical properties. The first ionization energy increases

▲ The general trend is of increasing first ionization energies across period 3 of the periodic table

across the period and peaks at a maximum with the halogens. The halogens therefore have the greatest effective nuclear charge of their respective periods, and this effective charge increases as you move **up** the group as the charge of the nucleus is less shielded. Fluorine has the highest effective nuclear charge of group 17 and will strongly hold on to its valence electrons. However, fluorine is also highly reactive as the high effective charge of its nucleus means flourine will ionize (remove electrons from) other species easily.

GENERAL

What are the trends in the chemical properties of elements in group 1?

The group 1 alkali metals all have one valence electron and form ionic compounds upon the formation of a positively charged cation, for example, Na^+ or K^+. The elements are being oxidized (see Chapter 10, Movement). We have seen that as you descend group 1, the ionization energy decreases due to an increase in the level shielding by the full inner electron orbitals. Less energy is required to remove the outermost valence electron and therefore the reactivity of the alkali metals increases as you move down the group.

Group 1 alkali metals react with water to form a metal hydroxide solution and hydrogen gas (see Chapter 2, Evidence). The metal hydroxide creates an alkaline solution. For example:

$$2Na(s) + 2H_2O(l) \rightarrow 2NaOH(aq) + H_2(g)$$

As you move down the group, the reactivity increases. For example, lithium reacts slowly but potassium reacts violently and produces enough heat to ignite the hydrogen gas.

1. Hypothesize whether the alkali metals will follow the same trend in reactivity when they react with oxygen.

GENERAL

What are the trends in the chemical properties of elements in group 17?

Group 17, the halogens, are non-metals characterized by seven valence electrons and a tendency to gain a single electron to attain a noble gas electron configuration. They form ionic compounds with metals, and with non-metals they form covalent compounds (see Chapter 9, Models). At standard temperature and pressure, they exist as diatomic molecules.

Although the group 1 alkali metals follow characteristic patterns in reactivity, chemical reactions and state of matter, the halogens are unique within the periodic table, as their state of matter changes as you proceed down the group. Fluorine and chlorine are gases, bromine is a liquid, and iodine and astatine are solids.

Halogens are highly reactive, with their reactivity decreasing as you move down the group. We have already seen that non-metals "like" to gain electrons or be reduced. As you go down the group, the number of full inner electron shells increases and the effective

◁ Chlorine, bromine and
iodine exist as diatomic
molecules: Cl_2, Br_2, and I_2.
Chlorine is a gas at room
temperature; bromine
is a dark liquid at room
temperature, although it
readily produces a brown
vapor and iodine is a
crystalline solid

nuclear charge decreases, due to an increase in the shielding of the nucleus by full electron shells. The result of this is that it is harder for elements lower in the group to obtain the extra electron they require to fill the outermost electron shell.

It's important to remember that compounds have very different properties to their constituent elements. A good example of this is sodium chloride, an ionic salt. Sodium is highly reactive with water and oxygen in air and must be stored under oil for safety. Chlorine is a yellow-green gas that is highly poisonous to humans when inhaled. The reaction between sodium and chlorine creates the salt, sodium chloride. The reaction is very exothermic producing a lot of heat and light energy:

$$2Na(s) + Cl_2(g) \rightarrow 2NaCl(s)$$

1. Compare the chemical and physical properties of sodium and chlorine with those of sodium chloride.

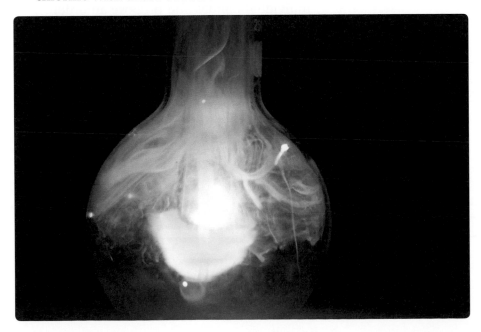

◁ When a piece of heated
sodium is lowered into a
flask of chlorine gas, the
reaction is violent and
exothermic

Experiment

Reactions between halogens and halides

We have established that as you move down group 17, the reactivity of the halogens and their non-metallic characteristics decrease as their effective nuclear charge decreases. We can investigate the reactivity of the halogens by performing a series of competitive reactions between halogens and their metal halides. These are displacement reactions and have the following general equation:

$$X_2(aq) + 2MY(aq) \rightarrow 2MX(aq) + Y_2(aq)$$

⚠️ Safety

- Wear safety glasses.

- Avoid breathing in fumes. Take special care if you are asthmatic.

- Keep the laboratory well ventilated.

- Dispose of waste solutions appropriately.

Materials

- 1.0 mol dm^{-3} solutions of potassium chloride, potassium bromide and potassium iodide

- Chlorine water (diluted 50% with water)

- 0.002 M bromine water

- Iodine water

- Six boiling tubes

- Boiling tube rack

- Six pasteur pipettes

Method

1. Using three separate pasteur pipettes, draw up a small amount of chlorine, bromine and iodine water. Record the color in your notes. You will use this information from these observations later in the experiment.

2. Place 5 cm^3 of chlorine water into a boiling tube.

3. To this boiling tube, add 5 cm^3 of potassium bromide solution. Mix the contents and in a copy of the table below, record any color changes that you observe.

4. Repeat steps 2 and 3 for the remaining combinations in the table.

	KCl(aq)	KBr(aq)	KI(aq)
Cl$_2$(aq)	no reaction		
Br$_2$(aq)		no reaction	
I$_2$(aq)			no reaction

Questions

1. After discussing your results with other students, explain why there is no reaction between chlorine water and potassium chloride.

2. Based on your initial observations of the color of chlorine, bromine and iodine water, draw conclusions about the color changes you observed in each of the individual reactions.

3. Construct balanced equations for the reactions which occurred during the experiment.

4. Analyze the pattern in the reactivity of the individual halides and propose an order of reactivity for the halogens based on experimental evidence.

Extension

5. Spectator ions are unchanged on both the reactant and product side of the chemical equation. Remove the spectator ions and write ionic equations for each of the reactions.

Do the elements of a period exhibit a pattern of acid–base properties?

An oxide is formed when an element and oxygen combine. The oxides of alkali metals and alkaline earth metals react with water to form metal hydroxides. The solution that results from this reaction is alkaline as it has a pH greater than 7. Therefore, metal oxides are defined as being basic. The reactions of sodium oxide and calcium oxide with water are described by the following equations:

$$Na_2O(s) + H_2O(l) \rightarrow 2NaOH(aq)$$

$$CaO(s) + H_2O(l) \rightarrow Ca(OH)_2(aq)$$

There are several different definitions of an acid and a base. One definition, the result of the collaborative work of Brønsted and Lowry (see Chapter 3, Consequences), refers to the hydrogen ion, H^+, as a proton, and defines an acid as a proton donor and a base as a proton acceptor.

A chemical species that behaves as both an acid and a base is called an amphoteric substance. These species can both donate and accept a hydrogen ion. An example of an amphoteric substance is water. A water molecule can either accept a hydrogen ion to form a hydronium ion H_3O^+ or donate a hydrogen ion to form a hydroxide ion OH^-.

$$H_2O + H_2O \rightleftharpoons H_3O^+ + OH^-$$

Aluminium oxide is an amphoteric oxide as it can react both as an acid and as a base. The chemical reactions of aluminium oxide are relatively complex. You will learn more about them if you go on to study chemistry at a higher level, but for now it is sufficient to be aware of its amphoteric nature.

The reactions of non-metal oxides with water create acidic solutions. For example, carbon, sulfur and phosphorus oxides react with water to create acidic solutions of varying strength. The balanced chemical equations for these reactions are as follows:

$$CO_2(g) + H_2O(l) \rightleftharpoons H_2CO_3(aq) \qquad \text{(carbonic acid)}$$

$$2NO_2(g) + H_2O(l) \rightleftharpoons HNO_3(aq) + HNO_2(aq) \qquad \text{(nitric acid and nitrous acid)}$$

$$SO_3(l) + H_2O(l) \rightleftharpoons H_2SO_4(aq) \qquad \text{(sulfuric acid)}$$

$$SO_2(g) + H_2O(l) \rightleftharpoons H_2SO_3(aq) \qquad \text{(sulfurous acid)}$$

$$P_4O_{10}(s) + 6H_2O(l) \rightleftharpoons 4H_3PO_4(aq) \qquad \text{(phosphoric acid)}$$

(Note: you are not expected to remember these equations.)

1. Consider the reactions of the following Group 3 oxides with water: sodium oxide, aluminium oxide, and sulfur dioxide. Describe whether each is acting as an acid or base. What does this suggest about the change in acid–base properties across a period?

In Chapter 3, Consequences, we looked at the global problem of acid deposition or acid rain. To recap, a product of the industrialization of developing economies, the rapid increase in acid-forming pollutants such as nitrogen and sulfur oxides can result in acid rain. Rainwater is slightly acidic due to the presence of the weak acid carbonic acid. It has a pH of approximately 5.6. Nitrogen oxides are the product of natural occurrences such as volcanic eruptions. The combustion of fossil fuels rich in sulfur impurities produces large amounts of acidic sulfur oxides.

Demonstration

The reaction of metal and non-metal oxides and water

⚠ **Safety**

- Appropriate safety glasses should be worn by the teacher and students.

- Demonstration should be performed in a fume hood. If this is not available, do it in a well-ventilated room behind a safety screen.

Materials

- Five gas jars and lids
- Oxygen gas cylinder*
- Deflagrating spoons
- Sodium metal
- Magnesium ribbon; do not look directly at burning magnesium
- Steel wool (iron)

- Sulfur powder; beware of toxic fumes
- Wood charcoal
- Universal indicator (UI) solution
- Distilled water bottles
- 50 cm³ beaker
- 200 cm³ beaker

*If an oxygen gas cylinder is unavailable, you can generate oxygen gas using the following chemicals and apparatus setup:

- Water trough and beehive shelf
- Flat bottom flask
- Rubber band and thistle funnel

- 100 vol hydrogen peroxide solution
- 1 g granular/0.1 g powdered manganese(IV) oxide catalyst

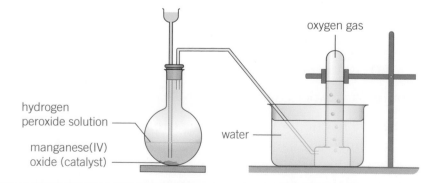

Method

1a. Using an oxygen gas cylinder, fill five gas jars with oxygen, carefully sealing each jar with a ground glass lid.

1b. If you are not using an oxygen gas cylinder, add water down the thistle funnel until the bottom of the funnel is below water level then add hydrogen peroxide.

2. Light a Bunsen burner and place a small piece of charcoal on a deflagrating spoon.

3. Strongly heat it in the Bunsen burner flame until ignition occurs. Remove the lid of a gas jar and quickly place the deflagrating spoon into the oxygen-filled gas jar.

4. Record your observations. When the reaction has finished, remove the deflagrating spoon and seal the gas jar with its lid. Place the gas jar to one side and retain for further analysis.

5. Repeat steps 2 to 4 for each of the remaining elements: sodium, magnesium ribbon, steel wool and sulfur.

6. Prepare a dilute solution of universal indicator by adding 12–18 drops of universal indicator liquid to $120\,cm^3$ of distilled water in a beaker. Keep a sample of this solution in a small beaker for comparison with the reaction jars.

7. While trying to limit the loss of the contents of the gas jar, quickly remove the lid and add approximately $20\,cm^3$ of the universal indicator solution to each jar, replacing the lid straightaway.

8. Mix the contents of each jar by shaking it vigorously.

9. Observe the color of the universal indicator solution in each jar and compare it to the color of the original UI solution.

Questions

1. Construct a table and record your observations. Include the following:

    a) the name and symbol of the reactants of each reaction

    b) a description of the reaction including the color of any flame produced when the reactant was heated

    c) the initial and final color

    d) pH of the final reaction mixture.

2. Using a universal indicator color strip, estimate the pH of each solution and record it in your data table.

3. Separate the reactants into two groups: metals and non-metals. Can you identify a trend in the pH values for each of these groups?

4. Write a general statement to describe the acid–base nature of metallic oxides and non-metallic oxides.

Summative assessment

Statement of inquiry:

Chemists look for patterns in the periodic table in order to discover relationships and trends that help them to predict physical and chemical properties.

Introduction

In this summative assessment, you will examine the periodic properties of elements. Then you will analyze an experimental procedure that examines patterns in the thermal conductivity of metals. You will then examine the work of pioneering chemist Döbereiner by analyzing the physical and chemical properties of the halogens. Finally, you will investigate the structural patterns created by water molecules in nature.

Patterns in physical and chemical properties

You are going to analyze and evaluate the trends in data for the first ionization energy (kJ mol⁻¹) of the first 20 elements of the periodic table.

Group								
1	2	3–12	13	14	15	16	17	18

H 131								He 2372
Li 520	Be 900		B 801	C 1086	N 1402	O 1314	F 1691	Ne 2081
Na 496	Mg 738		Al 578	Si 797	P 1012	S 1000	Cl 1251	Ar 1520
K 419	Ca 590							

▲ The first ionization energies (kJ mol⁻¹) of the first 20 elements

1. Identify elements with the lowest first ionization energy. What do you observe about the group they come from in the periodic table? [2]

2. Explain why the elements identified above tend to lose an electron. [3]

3. Identify the elements with the highest first ionization energy.

    a) Are they all members of the same group? [2]

    b) Explain why it is so difficult to remove an electron from these elements. [2]

4. **a)** State the trend in first ionization energies as you move down a group in the periodic table. [1]

   b) Suggest reasons why this trend in ionization energies exists. [3]

The following chart shows the melting and boiling points of the first 20 elements of the periodic table.

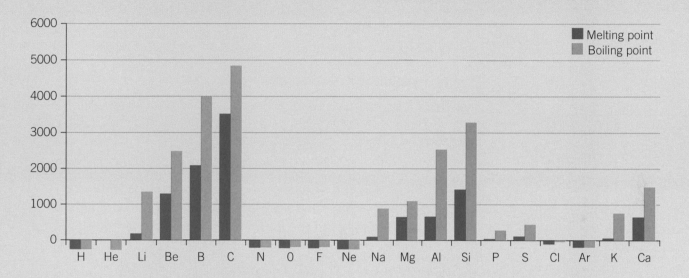

5. Examine the graph and describe the pattern in melting points as you move from left to right across period 3 (from sodium to argon). [3]

6. Chemists often refer to metals and non-metals when describing elements in the periodic table. Explain how you can identify metals and non-metals by examining patterns in the melting points of the individual elements. [3]

7. Which group of the periodic table has the maximum values for their melting points? Identify the type of bonding present in these elements and explain why the melting points are so high. [3]

Patterns in the thermal conductivity of a metal

Thermal conductivity is the rate of movement of heat through a material. By observing patterns in the way materials conduct heat, chemists have been able to classify specific groups of elements and compounds as high, medium and low thermal conductivity materials. Materials with low thermal conductivity are thermal insulators and materials with high thermal conductivities are used in applications that exploit this. In general, metallic materials have high electrical and thermal conductivity because of their metallic bonding.

The following is a method proposed by a student to test the thermal conductivity of a series of metals.

Materials

- 30 cm rods of copper, steel (iron), aluminium, brass and zinc
- Bunsen burner
- Retort stand and cork-lined clamp
- Stopwatch
- Wax candle and thumbtack

Method

- Secure a clamp to the end of one of the metal rods.
- Fix the clamp in place on the retort stand.
- Light a candle and drop some liquid wax onto the rod near to the clamp. Press the base of the thumbtack into the liquid wax and hold it until the liquid wax cools and the thumbtack is secured.
- Position the Bunsen burner at the opposite end to the thumbtack.
- Measure and record the distance between the thumbtack and the Bunsen burner.
- Light the Bunsen burner, adjust the flame to maximum and start the stopwatch.
- When the wax melts and the thumbtack falls from the metal rod, stop the stopwatch.
- Record your qualitative and quantitative observations.
- Repeat the procedure for each of the remaining metals.

8. Consider the method outlined above and suggest a possible hypothesis that the student was exploring. [2]

9. Evaluate the method and identify the independent, dependent and control variables. [3]

10. Discuss whether sufficient relevant data will be generated by this method. [2]

11. Explain how the method could be improved or extended. [3]

 ## Analyzing the periodic properties of the halogens

The German scientist Döbereiner developed the law of triads during the 1820s. His experiments and observations suggested that the chemical and physical properties of bromine were intermediate to those of chlorine and iodine; which we now know are, respectively, positioned above and below bromine in the periodic table. A lack of

experimental evidence about different triads within the periodic table meant that his findings were regarded by the scientific community of the time as interesting rather than groundbreaking.

A student performs a series of experiments to investigate the periodic properties of group 17, the halogens. The results and observations from the experiment are presented below.

	Chlorine	Bromine	Iodine
Boiling point (°C)	−34.04	58.78	184.4
State of matter at 25 °C	gas	liquid	solid
Appearance	green gas	brown liquid	black solid
Reaction with sodium (Na)	violent reaction, white crystalline (sodium chloride) solid that is soluble in water	vigorous reaction, white crystalline (sodium bromide) solid that is soluble in water	slow reaction, white crystalline (sodium iodide) solid that is soluble in water
Reaction with hot iron (Fe)	vigorous reaction iron(III) chloride formed Fe is oxidized	less vigorous reaction iron(III) bromide formed Fe is oxidized	mild reaction iron(II) iodide formed Fe is oxidized
Reaction with silver ions (Ag⁺(aq))	white precipitate	cream precipitate	yellow precipitate
Reaction with hydrogen gas (H₂)	formation of HCl explosive if exposed to flame	formation of HBr partially explosive if exposed to flame	formation of HI non-volatile.

12. Interpret the data and use your knowledge of patterns in physical and chemical properties to list arguments for placing these elements in the same group in the periodic table. [5]

13. Explain how the student could extend or improve this investigation into the halogens. [3]

 Patterns in nature

Snowflakes have fascinating structures. They can be found in numerous shapes, both regular and irregular. Snowflakes form in the clouds of water vapor, $H_2O(g)$. Once the temperature falls to 0°C or colder, the water undergoes a phase change from liquid to solid (ice). Scientists now know that the shape of the snowflake is affected by the temperature at which it forms, the presence or absence of dust particles, the relative humidity, air currents and the height of the clouds in which the crystals are formed. As a result, it is thought that there may be hundreds of different crystalline formations. The reason crystals grow in different ways remains a mystery despite decades of research and study—it is a scientific puzzle. However, the patterns formed are a thing of beauty.

The morphology of snow crystals was first described and classified in the Nakaya diagram, named after the Japanese physicist Ukichiro Nakaya, who grew snow crystals in his laboratory in the 1930s. Note that morphology is the study of form and structure.

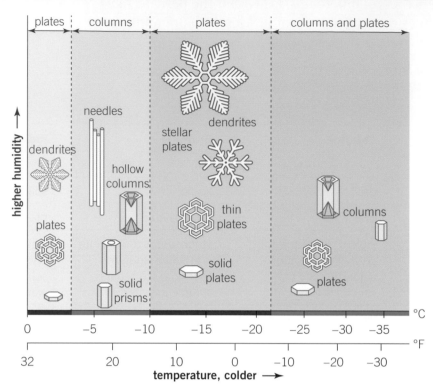

plates | columns | plates | columns and plates

higher humidity →

needles

dendrites

stellar
plates

dendrites

hollow
columns

columns

thin
plates

plates

solid
prisms

solid
plates

plates

°C

0   −5   −10   −15   −20   −25   −30   −35

°F

32   20   10   0   −10   −20   −30

temperature, colder →

▲ Nakaya diagram showing the formation of snow crystals

His diagram categorizes and describes the patterns formed by various snow crystal structures and identifies the effects of humidity and temperature on their formation. He subdivided falling snow into 41 morphological types. Researchers Magono and Lee classified 80 types in their 1966 scientific paper, and contemporary scientists hypothesize that there are hundreds of structures.

A snow crystal begins as a minute droplet which freezes into ice when the temperature is sufficiently cold. The ice particle develops facets becoming a small, hexagonal plate. This six-sided symmetry arises from the arrangement of water molecules in the ice crystal. Six branches emerge from the corners and the crystal grows larger. The molecular structure that occurs at the crystal's surface as a result of the range of conditions described earlier, and the exact path the crystal takes through the clouds, all contribute to determining its final shape. Because of the endless variables, no two crystals are shaped exactly alike. By studying the growth of snow crystals, we can understand more generally how molecules condense to form crystals, and this has applications in many scientific and industrial areas, both existing and yet to be discovered.

With reference to the Nakaya diagram, consider the following questions.

14. Identify the conditions necessary for large dendrite plate-shaped crystals to form. You should state both the temperature range and hollow columns. [2]

15. Identify and state the temperature at which needle-shaped crystals form. [1]

16. Which requires more humidity to form: solid plates or hollow columns? [2]

17. Suggest reasons why the work of Nakaya was so important to the field of crystal research. [3]

18. Create a detailed and well-labelled diagram, flowchart or other visual means to explain the formation of a snow crystal. [6]

12 Transfer

Many species of salmon are born in a freshwater environment, in streams and rivers far inland from the waters of the Pacific and northern Atlantic oceans where they live and grow. On their return to the freshwater to spawn, salmon transfer large amount of biomass from the oceans to inland aquatic systems. After spawning most of the salmon die, providing valuable food for a wide number of species such as eagles, bears, otters and predatory fish. What impact do you think salmon farming may have on an ecosystem?

"Food miles" is a term that refers to the transportation distance travelled by food from where it was produced to where it is consumed. Traditionally, fruit and vegetables were only available at certain times of the year when they were "in season" in a country. Nowadays, fresh green beans from Thailand, bananas from the Philippines, oranges from Brazil, blueberries from the United States and sweet potatoes from Uganda are available in markets throughout the world at all times. Does your family eat produce from other countries when they are "out of season" in your home country?

With over 3.5 million people passing through the station each day, Shinjuku train station in Tokyo, Japan is the busiest transport hub in the world. The transfer of people from point to point is a vital service provided by the state and private companies for people living in and around super-cities. How do scientific and technological advances in transport and infrastructure impact the communities and environments of satellite towns?

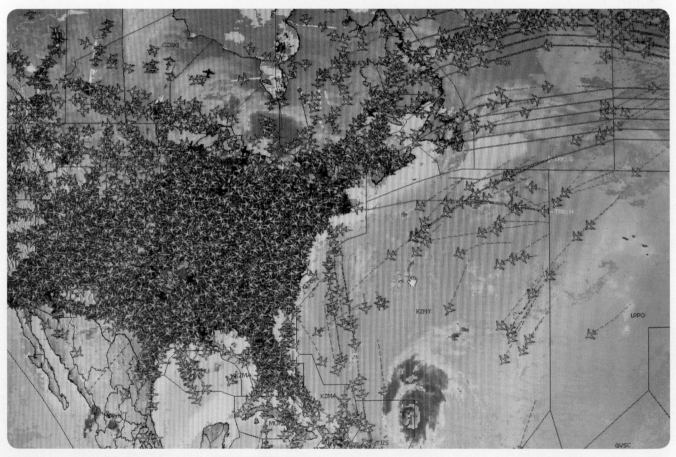

▲ Air travel has become increasingly affordable, and the number of planes transporting people around the globe has increased yearly. The International Air Transport Association (IATA) estimates that there are more than 100,000 flights daily carrying over 8 million passengers. Over 50 million tonnes of cargo is transported each year and the aviation industry supports 57 million jobs. Flying remains one of the safest modes of transport. In what ways have scientists developed innovations to reduce the damaging impact of air travel on the environment?

Key concept: Change

Related concept: Transfer

Global context: Scientific and technical innovation

Introduction

Transfer is the net movement of matter or particles from one location to another within a system. The law of conservation of matter is used in many fields of science. In biology, matter is transferred throughout an ecosystem because of the interactions between plants, animals and the physical environment. From a biological context, most systems are open, with matter and energy being transferred into and out of the system. While matter is free to move from one location to another, the amount of matter is conserved. In physics, scientists examine the transfer of matter and subsequent collisions at speeds close to the speed of light with the aim of gaining a better understanding of the fundamental laws of nature.

▼ In Italy, characteristic red soils are formed around a freshwater lake in a former quarry where bauxite was extracted. Aluminium is transferred out of bauxite for a variety of industrial applications. What are some of these applications?

Statement of inquiry:

Technological advances in analytical devices enhance the ability of scientists to monitor the transfer of matter when changes occur during chemical reactions.

Matter can be defined in many ways, but a general definition is anything that has a mass and occupies a space within the universe. Its transfer from place to place is an inherent part of every chemical reaction. Examples of this include:

- the movement of electrons during the breaking and making of chemical bonds in a chemical reaction

- the passage of ions through an electrolytic solution in a voltaic cell, resulting in the conversion of chemical energy into electrical energy

- the combination of acids and bases in neutralization reactions

- increasing or decreasing levels of vibrations of atoms during a change of state

- the transformation of matter from raw materials into products during industrial processes.

When chemists are at work in a laboratory, they constantly make observations using their senses or instrumentation. Data collected is then analyzed to establish if trends and patterns exist. Repeated trials that generate a large number of measurements is a protocol used to improve the reliability of data collected. Improvements in technology have enabled chemists to produce large volumes of data accurately and efficiently.

Scientists recognize that there is no one scientific method; they use a wide range of methodologies to generate data to understand how matter changes during a chemical process. Nature.com defines analytical chemistry as "a branch of chemistry that deals with the separation, identification and quantification of chemical compounds".

The industrial synthesis of compounds ranges from high-volume, low-value catalysed reactions to low volume, high-value specialist manufacturing. The transfer of matter from reactants to products in industrial processes and all forms of synthesis requires the identification of products, to confirm composition. This is a qualitative process. Measuring the purity and concentration of products is a quantitative process.

For these reasons, the concepts of change and transfer, and the global context of scientific and technical innovation will be examined.

▲ Zinc is used as a sacrificial anode, being preferentially oxidized while protecting the iron pipe from oxidation or rusting. Rusting is an electrochemical process where electrons are transferred through oxidation and reduction. Where is zinc positioned relative to iron in the activity series if it is preferentially oxidized?

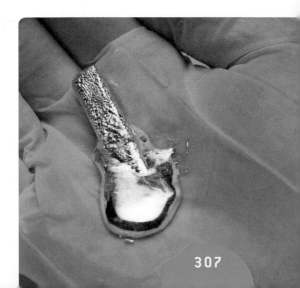

▶ Metallic gallium does not occur in nature in an uncombined state. Instead it is found in gallium(III) compounds within bauxite and zinc ores. With a melting point close to room temperature, the melting point of this metal can be exceeded by the warmth of the palm of your hand. Which other metal is a liquid at room temperature?

What is concentration?

The mole is an SI unit (Système International d'Unités), defined as a fixed amount of a substance (see Chapter 7, Function). A mole of any substance always contains the same number of particles, equal to Avogadro's constant (6.02×10^{23}). Most commonly we use this form of measurement when we examine reacting quantities in a chemical reaction.

The transfer of matter between reactants to give the products is the subject of much investigation in the scientific world. Stoichiometry examines the relationships between the amounts of reactants and products in a chemical reaction. Breaking the bonds of the reactants is an endothermic process and making new bonds to form the products is an exothermic process, so transfers of energy between the system and the surroundings are also required. Controlling the amount of the reactants, products and the conditions so that a reaction can occur is at the centre of the work of chemists and chemical engineers.

A solution is a homogeneous mixture of a solute—a solid—which has been dissolved in a solvent—a liquid (see Chapter 6, Form). For example, if you make a cup of hot chocolate, the solvent is warm milk and the solute is cocoa powder. Mixed well, it forms a homogeneous mixture. The addition of marshmallows creates a heterogeneous mixture as the composition of the mixture is no longer consistent. As the marshmallows are transferred from one place to another in the hot chocolate, the composition changes. Chemical systems are not unlike this example. When you prepare a solution, solids are transferred into the aqueous solvent, such as with the addition of copper(II) sulfate crystals into distilled water. When completing this task, the method will always instruct you to stir the mixture. Why is this?

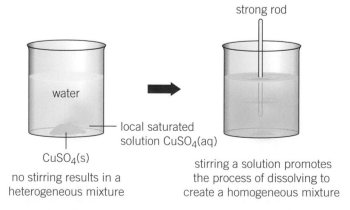

strong rod

water

local saturated solution $CuSO_4(aq)$

$CuSO_4(s)$

no stirring results in a heterogeneous mixture

stirring a solution promotes the process of dissolving to create a homogeneous mixture

▲ Stirring a mixture helps a solute to dissolve evenly in a solution

Many of the chemical reactions that you have already performed will have involved aqueous solutions. You will have heard terms such as "concentrated" and "dilute"; these are qualitative descriptions of the amount of matter that is dissolved in a measured quantity of a solvent.

The concentration of a solution can be described quantitatively through the calculation of the amount of matter in a given volume of solution. Qualitative observations can also give indications about concentration; for example, differences in the color of the metal salt cobalt(II) chloride are an indication that the solutions vary in their concentration.

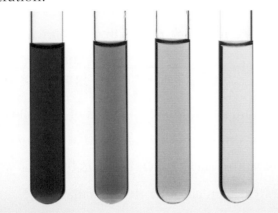

◀ Solutions of cobalt(II) chloride in water with the concentration decreasing from left to right

How can we calculate the concentration of a solution?

QUANTITATIVE

When we perform a chemical reaction we take great care to accurately measure the mass of a reacting solid, especially if the compound is expensive or is the limiting reagent within the reaction. The same care is required when dealing with aqueous solutions of chemicals. A solution can be regarded as a different way of transferring or delivering a specific substance, the solute, to the reaction mixture.

For example, consider the reaction between silver nitrate and sodium chloride solutions:

$$AgNO_3(aq) + NaCl(aq) \rightarrow AgCl(s) + NaNO_3(aq)$$

To better understand this reaction, we can split the ionic salts into their ions and identify the spectator ions which are not taking part in the precipitation reaction to form solid silver chloride.

$$Ag^+(aq) + NO_3^-(aq) + Na^+(aq) + Cl^-(aq) \rightarrow AgCl(s) + Na^+(aq) + NO_3^-(aq)$$

Now remove the spectator ions, $Na^+(aq)$ and $NO_3^-(aq)$, from the equation to leave:

$$Ag^+(aq) + Cl^-(aq) \rightarrow AgCl(s)$$

The water does not play a role in this reaction; the dissolved silver and chloride ions are taking part in it. It is these ions whose transfer into the system we need to control accurately.

To calculate the concentration of a solution, we measure the amount in moles (mol) of solute and the volume of the solvent in cubic decimetres (dm^3):

$$\text{concentration (mol dm}^{-3}) = \frac{\text{amount of solute (mol)}}{\text{volume of solution (dm}^3)}$$

(Remember $1\,dm^3 = 1\,litre = 1,000\,cm^3$.)

We can calculate the amount (in mol) of a substance using the following formula (see Chapter 7, Function):

$$\text{amount of substance (mol)} = \frac{\text{mass (g)}}{\text{molar mass } M_r \text{ (g mol}^{-1})}$$

Combining these two formulae results in the following:

$$\text{concentration (mol dm}^{-3}) = \frac{\text{mass (g)}}{\text{molar mass } M_r \text{ (g mol}^{-1}) \times \text{volume (dm}^3)}$$

1. Calculate the amount (in moles) in each of the following masses:

   a) 4.00 g of sodium hydroxide NaOH ($M_r = 40.00$ g mol^{-1})

   b) 58.80 g of sulfuric acid H_2SO_4 ($M_r = 98.09$ g mol^{-1})

   c) 14.6 g of hydrogen chloride HCl (calculate M_r)*

   d) 58.8 g of nitric acid HNO_3 (calculate M_r).*

   *Use the values of A_r on the periodic table.

2. Calculate the concentration, in mol dm^{-3}, of the following:

   a) 0.20 mol of nitric acid, HNO_3, dissolved in water to make a 0.50 dm^3 solution

   b) 8.00 g of sodium hydroxide, NaOH, dissolved in water to make a 0.80 dm^3 solution

   c) 0.45 mol of sulfuric acid, H_2SO_4, dissolved in water to make a 0.75 dm^3 solution

   d) 6.80 g of ammonia, NH_3, dissolved in water to make an 800 cm^3 solution.

QUANTITATIVE

Which is more important: accuracy or precision?

When you carry out an experiment, there are uncertainties linked to any measurement. The quantitative data that you record will probably be inexact. The extent to which it is inaccurate depends on the type of apparatus used.

It is important to understand the difference between accuracy and precision.

● Accuracy is the closeness of the agreement between a measurement and the true value of the quantity being measured.

- Precision is a comparison of how close a series of measurements are to one another; how close they are to the actual true value is not relevant.

1. Two students are each asked to measure out three 9.5 g samples of a reactant. When their teacher checked their results, she found the following:

	Student A	Student B
Sample 1:	12.2 g	11.6 g
Sample 2:	10.9 g	12.8 g
Sample 3:	6.4 g	12.1 g

Comment on the level of accuracy and precision of both sets of data.

2. The difference between precision and accuracy is important in the chemical laboratory. When you collect data, which of the following situations is easier to explain and why: results with poor precision or results with poor accuracy?

not accurate
not precise

accurate
not precise

not accurate
precise

accurate
precise

▲ Accuracy and precision are not the same thing

Data-based question: Precision of laboratory apparatus

Preparing solutions with accurate concentrations requires the use of laboratory apparatus that can measure volumes accurately. Apparatus such as a measuring cylinder are not sufficiently accurate to do this. Equipment such as burettes, pipettes and volumetric flasks are required to achieve the necessary level of accuracy.

1. a) Copy the table below and complete the missing data.

   b) Which piece of apparatus listed in the table is the most accurate?

Apparatus	Volume (cm³)	Absolute uncertainty (cm³)	Uncertainty
Burette	50.00	±0.05	0.10%
Bulb pipette	25.00		0.24%
Graduated pipette	10.00	±0.05	0.50%
Volumetric flask	100.00	±0.08	
Measuring cylinder	50.00		2.0%

2. What are the advantages and disadvantages of laboratory apparatus with fixed volumes?

3. Suggest reasons why measuring a volume of a liquid with a measuring cylinder is not good laboratory technique.

How do significant figures affect calculations?

Assigning a level of uncertainty to measurement apparatus in a laboratory is a way of describing the precision of the data being collected. Precision is more important in some measurements than others.

- To measure a mass of sugar to be used in a simple solubility experiment a top-pan balance that can read to 0.01 g is sufficiently precise.

- To prepare a standard solution for a titration an analytical balance that reads to 0.001 g is required.

The number of significant figures recorded is one of the ways in which scientists communicate the precision of a measurement. The fewer the number of significant figures, the less precise the measurement and the greater its uncertainty, and that of any value calculated from it. Generally, the cost of scientific equipment increases significantly as the level of precision of the instrument increases. For example, the cost of a four-decimal place analytical balance that can weigh to the nearest 0.0001 g is too expensive for most schools to be able to purchase.

It is important to record data to the correct number of significant figures. If you weigh yourself on a set of digital bathroom scales, it would be misleading to report your weight as 65.540 kg if the digital reading was actually 65.54 kg. By adding the zero digit in the third decimal place, you are indicating that the measurement has certainty to three decimal places.

There are a few simple rules to assign the number of significant figures of a number.

- All non-zero integers **are** significant (1, 2, …8, 9 are significant).

- Any zeros before a non-zero digit **are not** significant (e.g. 0.032 has 2 significant figures).

- Any zeros after a non-zero digit **are** significant (e.g. 2005 has 4 significant figures; 0.330 has 3 significant figures).

- If the number is recorded with no decimal point, the zeros **are not** considered to be significant (e.g. 5000 has one significant figure; 5000.0 has 5 significant figures).

Numbers written using scientific notation should reflect the number of significant figures.

Measurement	Scientific notation	Number of significant figures
235 g	2.35×10^2 g	three
13.451	1.3451×10^1 g	five
0.00350 dm³	3.50×10^{-3} dm³	three
8.000 kg	8.000 kg	four
100 kg	1×10^2 kg	one

When multiplying or dividing measurements, the answer should be expressed to the same number of significant figures as the measurement with fewest significant figures. For example:

4.130 (4 s.f.) × 2.80 (3 s.f.) = 11.564 which should be written as 11.6 (3 s.f.).

1. Convert each of the following measurements into scientific notation and state the number of significant figures.

   a) 0.00045 mg

   b) 1545 kg

   c) 32.65 cm

   d) 12.056 pH units.

2. To how many significant figures should the answer of each of the following calculations be expressed?

   a) 3.864 × 52

   b) 3.24587 ÷ 13.5

What is the role of titration in investigating concentration?

ANALYTICAL

Volumetric analysis is a laboratory technique which involves two solutions. One of the solutions, called the standard solution, has a concentration which is known to a high degree of accuracy. The other solution has an unknown concentration. Titration is a technique used to determine this unknown concentration.

An acid–base titration is the most common form of this analytical technique. The standard solution or titrant can be either an acid or base solution. The chemical reaction that occurs between the acid and base is known as neutralization—it is an exothermic reaction. For example:

$$HCl(aq) + NaOH(aq) \rightarrow NaCl(aq) + H_2O(l)$$

hydrochloric acid + sodium hydroxide → sodium chloride + water

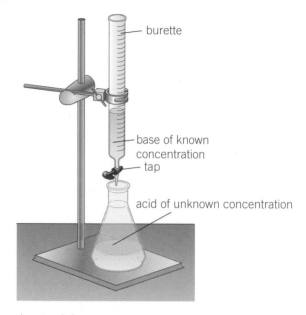

▲ Acid–base titrations are an accurate analytical technique for determining the concentration of a solution

The progress of the reaction is monitored by using an indicator. As the reaction approaches the equivalence point, a change in the color of the solution is observed. The equivalence point of this acid–base titration is the point when the exact same number of moles of hydrogen ions and moles of hydroxide ions have reacted.

An indicator is a chemical used to monitor the progress of a reaction involved in the transfer of liquids during a titration. Typically, it displays a different, distinct color depending on whether the environment is acidic or alkaline. It is normally either a weak acid or a weak base.

This concept can be examined by analyzing the chemical equilibrium that represents the action of an indicator as the pH of the reaction mixture changes:

$$HIn(aq) \rightleftharpoons H^+(aq) + In^-(aq)$$
$$\text{color A} \qquad\qquad \text{color B}$$

There are many acid–base indicators that can be used in titrations. The choice is dependent on the relative strength of the acid and the base involved in the neutralization reaction. For example, in an acid–base titration using the weak acid ethanoic acid and the strong base sodium hydroxide, the most suitable acid–base indicator for this titration is phenolphthalein.

To perform the calculations involved in an acid–base titration, there are several important steps:.

1. Write a balanced chemical equation for the reaction between the acid and the base.

2. Calculate the amount, in moles, of the solution of known concentration, the standard solution.

3. Use the mole ratio of the balanced chemical equation to determine the number of moles of substance present in the solution of unknown concentration.

4. Calculate the concentration of the solution.

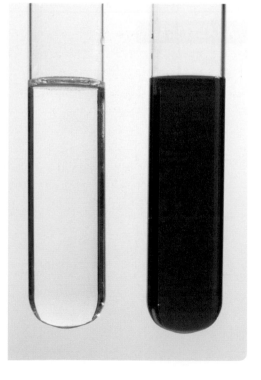

◄ The acid–base indicator phenolphthalein changes from colorless to light pink as the environment of the reaction mixture changes from acidic to alkali

Worked example: Acid–base titration calculations

Question

Calculate the concentration, in $mol\,dm^{-3}$ of $21.5\,cm^3$ potassium hydroxide solution that is neutralized by $25.0\,cm^3$ of $0.800\,mol\,dm^{-3}$ hydrochloric acid solution.

Answer

First, we need to write a balanced chemical equation.

hydrochloric acid + potassium hydroxide → potassium chloride + water

$$HCl(aq) + KOH(aq) \rightarrow KCl(aq) + H_2O(l)$$

Then we need to calculate the amount, in moles, of hydrochloric acid.

Convert the volumes of solution into dm^3:

$$\frac{25.0\ cm^3}{1000} = 0.0250\ dm^3$$

$$concentration\ (mol\,dm^{-3}) = \frac{amount\ of\ solute\ (mol)}{volume\ of\ solution\ (dm^3)}$$

Rearrange this equation so that the amount (in moles) is the subject of the equation.

$$amount\ of\ solute\ (mol) = concentration\ (mol\,dm^{-3}) \times volume\ of\ solution\ (dm^3)$$
$$= 0.800\ mol\,dm^{-3} \times 0.0250\ dm^3$$
$$= 0.0200\ mol$$

We then use the mole ratio of the balanced chemical equation to determine the number of moles of substance present in the solution of unknown concentration.

From the balanced chemical equation, the mole ratio between hydrochloric acid and potassium hydroxide is 1:1.

Therefore, 0.0200 mol of hydrochloric acid reacts with 0.0200 mol of potassium hydroxide.

Finally, we calculate the concentration of the potassium hydroxide solution.

$$concentration\ (mol\ dm^{-3}) = \frac{amount\ of\ solute\ (mol)}{volume\ of\ solution\ (dm^3)}$$

$$= \frac{0.0200\ mol}{0.0215\ dm^3} = 0.930\ mol\,dm^{-3}$$

 Experiment

An acid–base titration of a weak acid with a strong base

In this experiment you will determine the concentration of a weak acid using titration.

 Safety

- Wear safety glasses.

- Sodium hydroxide is a strong alkali and caustic in nature; take care when handling it.

Materials

- $0.50\,mol\,dm^{-3}$ solution of sodium hydroxide, NaOH (standard solution)

- Solution of ethanoic acid, CH_3COOH, of an unknown concentration (teacher to decide)

- Phenolphthalein indicator

- $50\,cm^3$ burette

- $25\,cm^3$ bulb pipette

←

- Pipette filler
- 150 cm³ conical flasks
- Retort stands and clamps
- White tile
- 250 cm³ beaker for waste solutions
- Plastic funnel

Method

1. Pipette 25 cm³ of ethanoic acid solution of known concentration into a clean 150 cm³ conical flask.

2. Add three or four drops of phenolphthalein indicator and swirl the flask to ensure the contents are mixed.

3. Prepare the sodium hydroxide in a burette.

4. Accurately record the initial volume of sodium hydroxide.

5. Titrate the ethanoic acid with the sodium hydroxide by opening the tap in the burette and allowing the alkali to slowly drop into the conical flask.

6. With each small addition of sodium hydroxide, swirl the flask.

7. As the point of neutralization approaches, you will observe a localized color change in the flask. Slow the addition of the sodium hydroxide to one drop at a time.

8. Observe and record the color change and the final volume of alkali required. This trial could be regarded as a rough titration.

9. Repeat the titration with a fresh 25 cm³ sample of ethanoic acid and a fresh conical flask.

10. Repeat the procedure until concordant results are obtained (that is, the volume of alkali added is within 0.1 cm³ at least twice).

11. Record all your qualitative and quantitative observations.

Questions

1. Describe the color change at the point of neutralization of the ethanoic acid by sodium hydroxide.

2. Write the balanced chemical equation for the reaction between ethanoic acid and sodium hydroxide.

3. Calculate the average volume (in dm³) of sodium hydroxide released from the burette.

4. Calculate the number of moles of sodium hydroxide in this volume.

5. Using the balanced chemical equation, calculate the concentration (in mol dm⁻³) of the ethanoic acid solution.

6. Compare this concentration with the actual concentration of ethanoic acid (your teacher will know this).

7. Comment on the accuracy of your result and your level of precision in using this analytical technique.

ACIDS AND BASES

How do pH curves help us to visualize the chemistry of an acid-base titration?

The transfer of acids and bases within a system can be monitored by analyzing changes in the pH of the mixture. The use of paper indicators such as litmus paper and indicator solutions such as methyl orange and phenolphthalein help us to observe the changes in pH as an acid–base reaction proceeds. However, these methods do not give us an accurate picture of the changes in pH over the course of the reaction.

Although we can observe changes in the color of indicators, to analyze the relationships and patterns between the reacting acids and bases, chemists use pH curves to track the change in pH of the neutralization reaction.

What does the shape of the pH curve tell us about an acid–base reaction?

The simplest pH curve is for the titration of a strong acid with a strong base, for example, the titration of hydrochloric acid, a strong acid, with sodium hydroxide, a strong base.

In this titration, the strong acid is placed into the conical flask, and the strong base is added from a burette into the conical flask. The transfer of an alkali of known concentration from the burette to the reaction mixture in the conical flask can be performed with great precision due to the accuracy with which this apparatus can measure a specific volume. The uncertainty in a measurement made by a standard $50 \, cm^3$ burette is $\pm 0.05 \, cm^3$.

1. Comment on what the starting point of the pH curve on the y-axis tells us about the strength of hydrochloric acid?

2. Interpret the pH curve and explain why the pH of the solution starts to increase as sodium hydroxide is transferred from the burette to the conical flask.

3. After $25 \, cm^3$ of sodium hydroxide has been transferred to the reaction mixture, there is a sudden increase in the pH level. What does this tell you about the amount of hydrochloric acid that remains unreacted in the conical flask? Justify your answer with scientific reasoning.

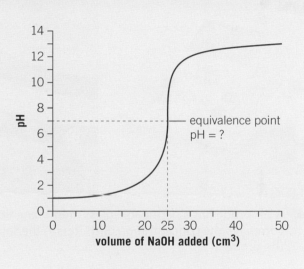

How can technology improve experimental measurements?

ANALYTICAL

Science provides knowledge and understanding of phenomena which enables us to make technological advances. In turn, technology can increase our ability to collect accurate data which we can use to test hypotheses and develop theories.

Modern research laboratories utilize technology to overcome the limitations of human senses in making observations during practical investigations. In school laboratories, we use many forms of technology too. For example, data probes can be used to collect and record a wide variety of information ranging from temperature, conductivity, pH, gas pressure, dissolved gases and color changes.

When linked to supporting software, the data can be analyzed and presented in a form that enables us to discover trends in the data and results.

For example, the pH probe exploits the presence of hydrogen ions throughout a solution. It has a permeable glass membrane through which the hydrogen ions can pass, but larger ions cannot pass. In simple terms, the pH probe acts as an electrochemical cell and measures the voltage, or potential difference of the solution. The relationship between the voltage and the concentration of hydrogen ions in solution is then determined mathematically.

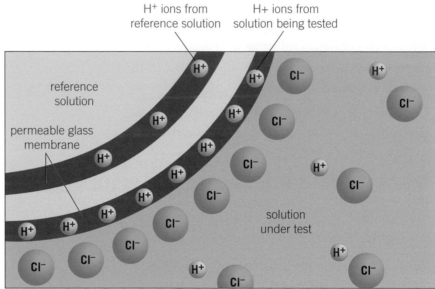

H⁺ ions from reference solution   H+ ions from solution being tested

reference solution

permeable glass membrane

solution under test

▲ A pH probe has a permeable glass membrane through which the hydrogen ions can pass. The probe compares the concentration of H⁺ ions in the solution being tested with a reference solution within the probe

What is the structure and nomenclature of a carboxylic acid?

Carboxylic acids are derived from alcohols and are formed by the oxidation of a primary alcohol.

▶ Ethanol, CH_3CH_2OH, is a primary alcohol containing the hydroxyl $-OH$ functional group

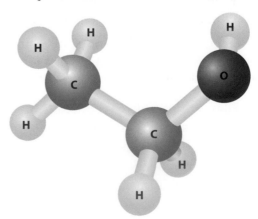

A primary alcohol has the hydroxyl (-OH) group bonded to a primary carbon atom in the compound (a primary carbon atom is bonded to only one other carbon atom). For the conversion of an alcohol to a carboxylic acid, an oxidizing agent, such as potassium manganate(VII), $KMnO_4$, needs to be used. However, this is not enough: for the reaction to proceed, the reaction mixture needs to be acidified and heated. In this process, the alcohol is first converted into an aldehyde and then into carboxylic acid, as shown in the diagram.

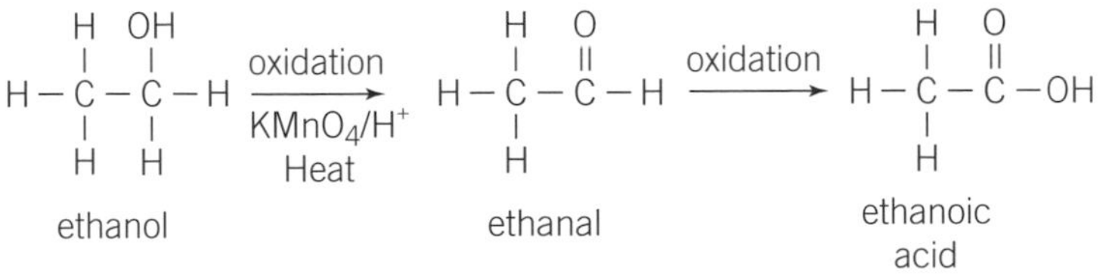

◀ Ethanol (an alcohol) can be oxidized to produce ethanal (an aldehyde) which is then oxidized to produce ethanoic acid (a carboxylic acid)

Naming carboxylic acids

The nomenclature rules for carboxylic acids closely follow those for naming alkanes (see Chapter 2, Evidence). It involves identifying the parent molecule and changing the suffix.

- The prefix or the first part of the carboxylic acid name is derived from the alcohol. This remains unchanged.

- The suffix or last part of the alcohol name is changed from -ol to -oic acid.

For example, when the alcohol propanol is oxidized, the carboxylic acid propanoic acid is formed.

1. State the name of the carboxylic acid formed when heptanol is oxidized.

How are esters manufactured?

ORGANIC

In nature, esters are found in animals and plants; they are responsible for the familiar fragrances given off by fruits. How do we detect these pleasant odours? When an organic compound is volatile, it tends to transform from a liquid to a gas. Diffusion is the transfer of matter, from one place to another. In the case of esters, the diffusion of a gas through the atmosphere allows us to detect smells with our senses. Esters have a wide variety of applications including as flavoring agents in the food industry.

Name of ester	Fruit	Structure
Pentyl ethanoate	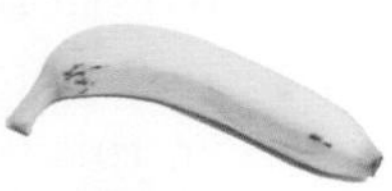	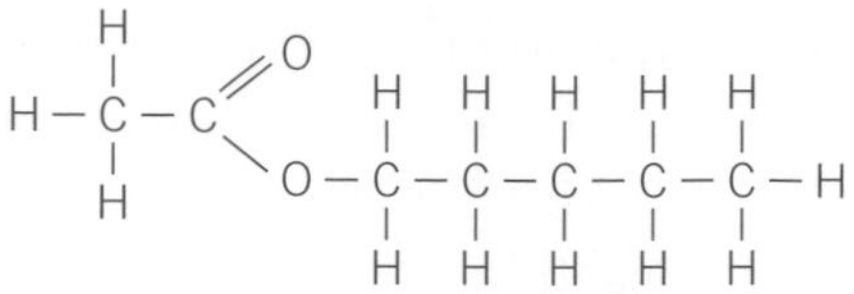
Ethyl butanoate		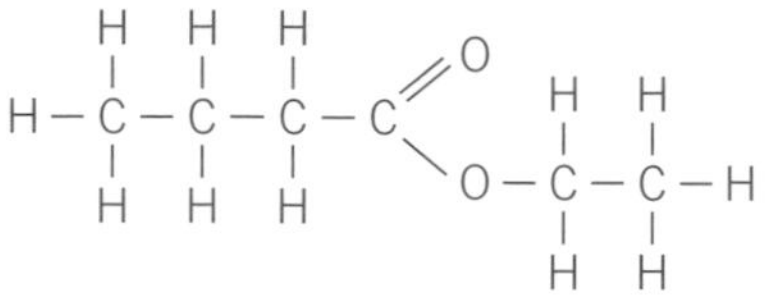
Methyl butanoate		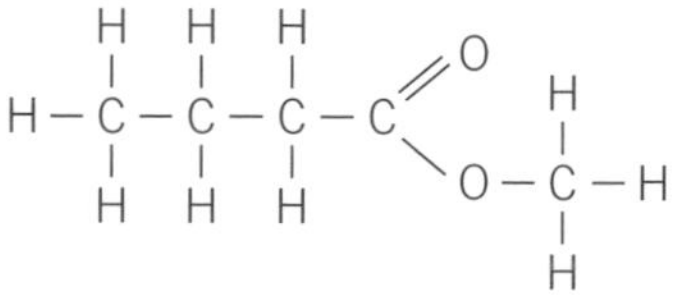
Pentyl butanoate		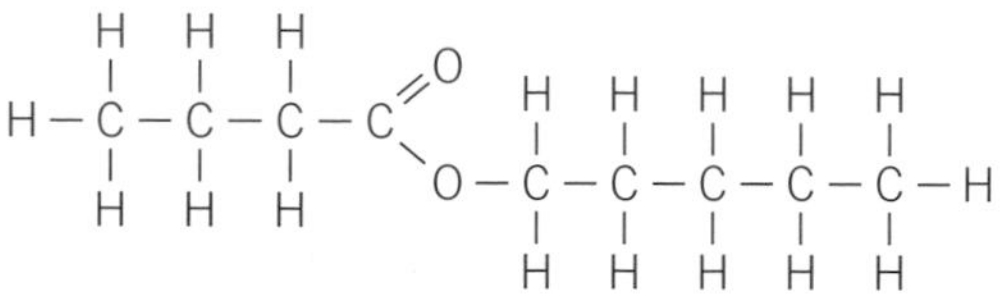

▲ Esters are small volatile organic compounds which are responsible for the characteristic fragrances of fruit

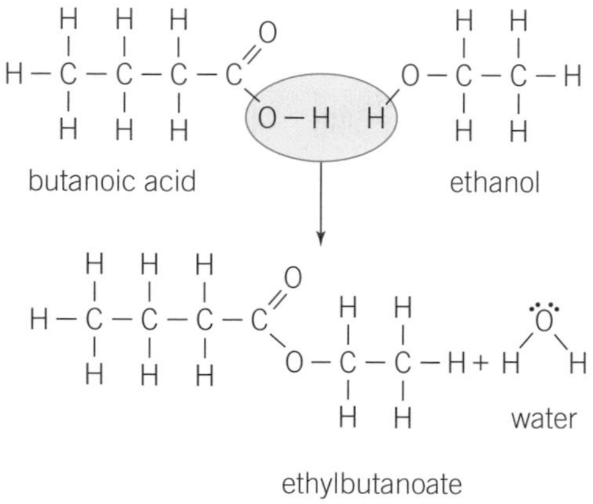

butanoic acid

ethanol

ethylbutanoate

water

Esters are formed by a condensation reaction between a carboxylic acid and an alcohol. A condensation reaction involves the combination of two different reactants which results in the removal of a water molecule. Ester formation is catalysed by concentrated sulfuric acid and the mixture must be heated.

Look at the word, chemical and structural equations for the formation of an ester:

$$\text{butanoic acid} + \text{ethanol} \rightarrow \text{ethyl butanoate} + \text{water}$$
$$C_4H_8O_2 + C_2H_6O \rightarrow C_6H_{12}O_2 + H_2O$$

You may also find it helpful to use molecular models to visualize the formation of esters.

1 How does using a structural formula help you to understand a chemical reaction? Identify the differences between structural and chemical formulae.

Naming esters

The nomenclature rules for esters involve identifying the reactants in the esterification process:

- The prefix or the first part of the ester name is derived from the alcohol.
- The suffix or last part of the alcohol name is changed from *-ane* to *-yl*.
- The second part of the ester name derives from the carboxylic acid.
- The suffix of the carboxylic acid name is changed from *-oic* to *-oate*.

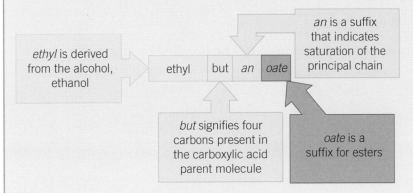

ethyl is derived from the alcohol, ethanol

ethyl | but | *an* | oate

an is a suffix that indicates saturation of the principal chain

but signifies four carbons present in the carboxylic acid parent molecule

oate is a suffix for esters

1. Name the ester which is produced from each of the following pairs of reactants:

    a) propanol and butanoic acid

    b) methanol and ethanoic acid

    c) hexanol and pentanoic acid

    d) butanol and methanoic acid.

ATL **Information literacy skills**

Finding, interpreting, judging and creating information

The ability to use the internet effectively to locate information, and then organize, analyze and evaluate it are essential skills of learning. It is also important to acknowledge the source and authorship of the materials. The presence of information on the internet is not a guarantee that the information is scientifically correct. What are the questions you can ask yourself to assess whether you are confident about the information it contains?

- Is the website that of a government department, a scientific research organization or an educational institution such as a university?
- Does the article acknowledge the author?
- Is the information on the website up to date?
- Does the article contain a list of references or in-text citations?

Use the internet to research and document some of the common and more unusual applications of the organic molecules called esters.

Experiment

Formation of a series of esters

Heating an alcohol and a carboxylic acid in the presence of concentrated sulfuric acid creates an ester.

⚠ Safety

- Wear safety glasses.
- The reaction must be performed in a fume hood.
- Concentrated sulfuric acid is highly corrosive and should not be handled by students.

Materials

- Boiling tubes
- $400\,cm^3$ beaker
- $100\,cm^3$ beaker
- Concentrated sulfuric acid
- Concentrated ethanoic acid
- $1.0\ mol\ dm^{-3}$ carboxylic acid solutions: ethanoic acid, propanoic acid, benzoic acid
- Alcohols: methanol, ethanol, propan-1-ol, butan-1-ol
- $1.0\ mol\ dm^{-3}$ solution of sodium carbonate solution
- Boiling water (heated in an electric kettle, not using a Bunsen burner)

Method

Use the method described here for each of the combinations listed in the table on the next page.

1. You will be provided with a stoppered test tube containing 6 drops of concentrated sulfuric acid.

2. Add 10 drops of the alcohol to this test tube.

3. Now add 10 drops of the carboxylic acid to the test tube containing the alcohol.

4. Using boiling water from an electric kettle, create a water bath in a $400\,cm^3$ beaker. (Do not use a Bunsen burner as alcohols are flammable.)

5. Place the test tube in the hot water bath and allow the mixture to heat for approximately 3–5 minutes.

6. Carefully pour the mixture into a small $100\,cm^3$ beaker containing $10\,cm^3$ of sodium carbonate solution. The role of the sodium carbonate solution is to neutralize any acid present in the reaction mixture. You will observe effervescence for a few minutes.

7. The ester should be found on the top of the solution. Waft the vapors towards your nose and record its odour in a copy of the table below.

Alcohol	Carboxylic acid	Description of odour detected	Name of ester produced
Methanol	Ethanoic acid		
Ethanol	Ethanoic acid		
Propan-1-ol	Ethanoic acid		
Butan-1-ol	Ethanoic acid		
Methanol	Propanoic acid		
Ethanol	Propanoic acid		
Propan-1-ol	Propanoic acid		
Butan-1-ol	Propanoic acid		
Methanol	Benzoic acid		
Ethanol	Benzoic acid		
Propan-1-ol	Benzoic acid		
Butan-1-ol	Benzoic acid		

Questions

1. Write the word equation for the reaction between methanol and ethanoic acid.

2. Deduce and record the name for each of the esters formed in your reactions.

3. Explain the purpose of the concentrated sulfuric acid in the esterification reaction.

Summative assessment

Statement of inquiry:

Technological advances in analytical devices enhance the ability of scientists to monitor the transfer of matter when changes occur during chemical reactions.

Introduction

In this summative assessment, we will first look at the role of indicators in acid–base titrations. We will then investigate the behavior of weak and strong acids, focusing on pH and concentration. Next we will process and examine the experimental data collected during an acid–base titration. Finally, we will examine the role of chemists in the modern food manufacturing industry.

 ## Analytical chemistry

Titration is a quantitative analytical technique that has been used by chemists in one form or another for over 150 years. Advances in technology have made data probes such as a pH probe reliable and more cost effective, and they are now commonly used in school laboratories. They can be used to track a change in pH over the course of a titration. Alternatively, indicators can be used to track pH changes visually.

The diagram shows the color changes of different acid–base indicators in acidic and alkaline environments.

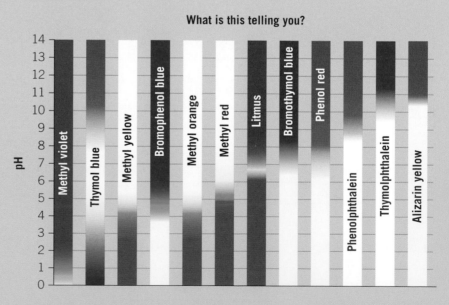

What is this telling you?

1. State the pH at which methyl red indicator changes color. [1]

2. List the indicators that have more than one distinct color change. [4]

3. Comment on the differences between phenolphthalein and bromothymol blue in terms of color changes. Identify and justify which indicator has a narrow pH range for the change in color. [4]

Examine the pH curves of two different acid–base titrations.

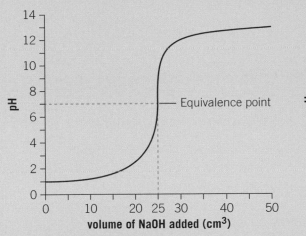

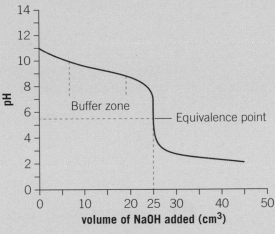

4. For the titration of a strong acid with a strong base, the equivalence point at which base completely neutralizes the acid has a pH of 7.0. Select and justify your choice of an indicator that will allow you to visually determine the equivalence point. [2]

5. Discuss why bromophenol blue would not be a good choice of indicator for the titration of a strong acid and a strong base. [2]

6. For the titration of a weak base with a strong acid, determine the initial color of solution if the indicator methyl yellow is used. State the color at the completion of the titration. [2]

7. Select and justify your choice of an indicator that will allow you to visually determine the equivalence point in the weak base–strong acid titration. [2]

 ## Investigating the transfer of hydrogen ions within an aqueous solution

We know from our work in Chapter 3, Consequences, that a strong acid will completely dissociate in solution into its ions, and a weak acid will undergo partial dissociation into its ions. The pH of a solution is a measure of the concentration of hydrogen ions, H^+, present in a solution. You are provided with the following materials:

- $0.10 \, mol \, dm^{-3}$ solution of ethanoic acid and hydrochloric acid

- $0.010 \, mol \, dm^{-3}$ solution of ethanoic acid and hydrochloric acid

- $0.0010 \, mol \, dm^{-3}$ solution of ethanoic acid and hydrochloric acid

- $0.00010 \, mol \, dm^{-3}$ solution of ethanoic acid and hydrochloric acid

- pH probes

8. Choose a hypothesis to investigate from the list below. Explain your choice with scientific reasoning. [4]

   a) A decrease in the concentration of an acid will increase the pH of the solution.

b) The degree of dissociation of an acid is dependent on the initial concentration of a weak acid.

c) There is no difference in the pH of a solution of a strong or weak acid, when the initial concentration is identical.

9. Design an experiment to test your hypothesis. Your method should include:

- the independent and dependent variables, and other variables being controlled

- how you will collect quantitative and qualitative observations in your investigation. [8]

 Analysis and evaluation

A student performed an acid–base titration in the school laboratory and collected quantitative data. The reaction was between a 0.10 mol dm⁻³ solution of sodium hydroxide NaOH, a strong base, and 25 cm³ of a weak acid of unknown concentration and composition. We can give this acid the general formula HA.

10. Write the chemical equation for the neutralization reaction between sodium hydroxide and the weak base HA. [3]

Using a pH probe, the following data was collected and tabulated.

Volume of NaOH added (cm³)	pH	Volume of NaOH added (cm³)	pH	Volume of NaOH added (cm³)	pH	Volume of NaOH added (cm³)	pH
1.00	2.95	8.00	3.85	15.00	5.15	18.80	11.20
2.00	3.06	9.00	4.00	16.00	5.41	19.00	11.40
3.00	3.21	10.00	4.21	17.00	5.81	19.20	11.55
4.00	3.35	11.00	4.330	18.00	6.75	19.50	11.65
5.00	3.44	12.00	4.50	18.20	10.5	20.00	11.71
6.00	3.59	13.00	4.82	18.40	10.75	21.50	11.80
7.00	3.7	14.00	4.98	18.60	11.05	22.00	11.85

11. Identify and explain which data was not correctly recorded. [2]

12. Analyze the graph and comment on how you think the accuracy of the data collected would or would not affect the shape of the curve. [3]

13. Display the data in a graph of pH versus volume of sodium hydroxide added. Connect the data points with a smooth curve. [4]

The point of neutralization for this reaction occurs when the number of moles of acid equals the number of moles of alkali. At this point, there will be a rapid change in the pH of the solution.

14. From your graph, estimate the volume of sodium hydroxide added to reach the point of neutralization. [1]

15. Calculate the number of moles of sodium hydroxide at the point of neutralization. [2]

16. Calculate the concentration (mol dm⁻³) of the weak acid. [2]

Researching the chemistry of flavors

This article is abridged from the website of the American Chemical Society. Read the passage below and answer the questions that follow.

The chemistry of flavors

Flavor – encompassing both aroma and taste – provides the defining characteristic of how we experience food. Flavor has long been an enigma to scientists: Aristotle described two categories of taste, sweet and bitter. Today we recognize five basic tastes in food: sweetness, saltiness, sourness, bitterness and umami (savory). But what are the scientific components of flavor, and how can flavor be studied, quantified and replicated?

Flavor is caused by receptors in the mouth and nose detecting chemicals found within food. These receptors respond by producing signals that are interpreted by the brain as sensations of taste and aroma. Certain taste and aroma combinations are characteristic of particular foods.

For example, a green apple tastes the way it does because the unique combination of chemicals found naturally within it are perceived by our mouths, noses and brains as the distinct blend of sweet and sour tastes and volatile aromas characteristic to the fruit. Identifying this chemical profile allows food producers to retain flavor in preserved green apples and, through synthesis of these flavor compounds, makes possible the production of candy, soda and other products using artificial green apple flavor.

The chemicals that produce flavors are notoriously difficult to study because a single natural flavor may contain hundreds or even thousands of component substances, and some of these substances are present in minute quantities. Understanding the components of flavor has become more important than ever with the modernization of food systems and the increased reliance on processed foods.

Why does a strawberry smell like a strawberry? To study its aroma, Dimick's team designed and built a system of separators and concentrators to obtain sufficient amounts of strawberry essence. They made use of evaporation, distillation and extraction technologies in their large and complicated device. After six seasons of processing 30 tons of fruit, they succeeded in obtaining a few grams of strawberry oil – the essence of strawberries that gives them their characteristic aroma.

17. From the article, identify the information required by chemists to synthetically reproduce natural flavors. [3]

18. Explain why the role of chemists in the food manufacturing industry is becoming increasingly important. [3]

19. Explain how information technology and instruments used in analytical chemistry are fundamentally important in the process of identifying the chemical components of natural flavors and the subsequent synthesis of artificial flavors. [3]

Glossary

Accuracy   is the degree to which a measurement represents the actual value of that which is being measured.

Acids   are chemical compounds which produce hydrogen ions $H^+(aq)$ when in aqueous solution, forming an acidic solution (pH < 7.0).

Activation energy   is the minimum amount of energy required to start a chemical reaction through initial bond breaking, allowing new bonds to be made.

Activity series   is a list of elements placed in order of their reactivity, as determined by their reaction with air (oxygen), water and dilute acid.

Alcohols   are organic compounds that contain the hydroxyl (OH^-) functional group.

Alkalis   are water-soluble bases which produce hydrated hydroxide ions $OH^-(aq)$ when in aqueous solution.

Alkali metals   are the elements in the group 1 of the periodic table, which all have a single valence electron. They form alkaline solutions (pH > 7.0) when they react with water.

Alkanes   are a homologous series of hydrocarbons with a general formula C_nH_{2n+2}.

Alkenes   are a homologous series of hydrocarbons with a general formula C_nH_{2n}.

Alloy   An alloy is a mixture of two or more metals, or a mixture of a metal and another element. Examples include steel, a mixture of iron and carbon, and brass, a mixture of copper and zinc.

Anhydrous   describes salts which have lost their water of crystallization.

Anions   are atoms or molecules containing more electrons than protons, so they carry a negative charge.

Aqueous solution   is a solution where the solvent is water.

Anode   an anode is an electrode in an electrochemical cell where the oxidation reaction takes place.

Atom   an atom is made up of protons, neutrons and electrons, and is the basic building block of all matter.

Atomic number   the atomic number is the number of protons an element has in the nucleus of its atom.

Avogadro's constant   is the number of representative particles in one mole of a substance. This number is 6.02×10^{23}.

Bases   are chemical compounds that react with acids to form a salt and water.

Biofuel	is plant material or animal waste which can be used as a fuel resource.
Boiling point	the boiling point is the temperature at which all of a liquid changes into a gas (or vapor). At this temperature, the vapor pressure of the liquid is equal to atmospheric pressure.
Brownian motion	is the random motion of particles in water or air caused by collisions with the surrounding molecules.
Burette	a burette is a graduated glass tube with a tap at one end, which is used in a titration to add controlled volumes of liquids with great accuracy and precision.
Catalyst	a catalyst is a substance that increases the rate of a chemical reaction without itself undergoing any permanent chemical change, by providing an alternate reaction pathway with a lower activation energy.
Cathode	a cathode is an electrode in an electrochemical cell where the reduction reaction takes place.
Cations	are atoms or molecules containing fewer electrons than protons, and so carry a positive charge.
CFCs	(abbreviation for chlorofluorocarbons) are inert chemicals which may be used as refrigerants or as solvents in aerosols.
Chemical change	a chemical change occurs during a chemical reaction when reactants are transformed into products. The new substances have different chemical and physical properties compared to the reactants. For example, when hydrogen burns in oxygen, water is formed. Water is a colorless liquid and has none of the properties of its constituent elements, which are both gases.
Chemical equilibrium	is a stage reached in a reversible chemical reaction when the forward and backward reactions take place at an equal rate.
Chemical formula	a chemical formula is a combination of symbols and numbers that represent the relative proportion of atoms of elements present in a chemical compound or molecule. For example, a water molecule, H_2O, is made up of two atoms of hydrogen and one atom of oxygen.
Chemical reaction	a chemical reaction is a change by which elements and compounds rearrange their atoms to produce new chemical elements or compounds.
Chromatography	is a physical separation technique where a mixture is carried by a solution, or in a gas stream, across an absorbent material. The components of the mixture are separated by this technique.
Closed system	a closed system is a system in which no matter can enter or leave the system, though energy can.
Collision theory	explains rates of reaction in terms of the motion of reactant particles.
Colloid	a colloid is a substance consisting of very small particles (about 10^{-4} to 10^{-6} mm across) suspended and dispersed in a medium such as air or water.
Combustion	is the chemical reaction of a substance with oxygen.

Compound a compound is the substance formed by the chemical combination of elements in fixed proportions, as represented by the compound's chemical formula.

Concentration is the amount (in mol) of a substance dissolved per unit volume (in dm^3) of a solvent (normally water).

Condensation is the change of state from a gas to a liquid.

Conductor a conductor is a material that allows the flow of energy, either in the form of electricity, thermal energy or sound energy. Electrical conductors allow the passage of electrons or charged ions.

Corrosion is a chemical reaction (oxidation) between a metal and oxygen gas found in air.

Covalent bond a covalent bond is formed by the electrostatic attraction between a shared pair of negatively charged electrons and the positively charged nuclei.

Covalent network solid A covalent network solid is a giant three-dimensional lattice structure in which atoms are held together by covalent bonds. Examples include diamond, graphite and silicon dioxide.

Decomposition is a chemical change by which a compound is broken down into simpler compounds or elements.

Deductive reasoning is a move in thinking from the general, in the form of a theory, to the specific, in the form of our observations.

Delocalized electrons are electrons that are spread over several atoms within a molecule or ion. They are not confined to one single atom.

Density is the mass per unit volume.

Diatomic molecule a diatomic molecule is formed from two atoms covalently bonded to each other. An example is hydrogen gas, H_2.

Diffusion is a physical occurrence where particles spread out from an area of high concentration to one of low concentration.

Dilute solutions have a low concentration. This means that the amount of solute for a given volume of solvent is small.

Dissociation is the process where acids and bases break up into their ions in solution. For example, the strong acid hydrochloric acid, HCl, completely dissociates into the hydrogen ion, H^+, and the chloride ion, Cl^-.

Distillation is a physical separation technique where volatile liquids are heated to their boiling point and the vapors are condensed and collected.

Double covalent bond a double covalent bond is the sharing of two pairs of electrons between two bonding atoms.

Ductility is the ability of a substance, normally a metal, to be drawn out into wire.

Electrochemical cell an electrochemical cell involves electrical energy–chemical energy interconversions. There are two main types of electrochemical cells: voltaic cells and electrolytic cells.

Electrode an electrode is a piece of metal or carbon (graphite) placed in an electrolyte which allows the passage of an electric current in an electrochemical cell.

Electrolysis	is a process that occurs in an electrolytic cell and involves the conversion of electrical energy into chemical energy.
Electrolytes	is the solution found in an electrochemical cell. It contains both anions and cations. These ions are involved in the oxidation and reduction reactions in these cells.
Electron	an electron is a negatively charged subatomic particle which is found in orbitals surrounding the nucleus of an atom.
Electron configuration	is the arrangement of electrons in the various electron shells around the nucleus of an atom.
Electronegativity	is the attraction an atom has for the shared pair of electrons in a covalent bond. Non-metals have higher electronegativity values than metals.
Electrostatics	the law of electrostatics states that like charges (two positive charges or two negative charges) repel one another, and unlike charges (positive and negative charges) attract each other.
Empirical data	is collected by means of experimentation and observations.
Empirical formula	an empirical formula is a chemical formula that shows the lowest whole number ratio between the atoms in a chemical formula.
Endothermic reaction	an endothermic reaction is a chemical reaction during which heat energy is absorbed from the surroundings.
Energy	is the capacity of a system to do work.
Energy density	is the amount of energy stored in a given mass of a substance.
Enthalpy	is the heat transferred by a system during a chemical reaction.
Entropy	is a measure of the distribution of available energy between particles. The greater the distribution or spreading of energy amongst the particles, the lower the chance that the particles will return to their original state. A gas has a higher entropy than a solid.
Equivalence point	is the point in an acid–base titration when the number of moles of solution of known concentration (titrant) added equals the number of moles of unknown solution (analyte).
Esters	are organic compounds formed by the reaction of a carboxylic acid and an alcohol. They are often sweet smelling liquids.
Evaporation	is the change of state from a liquid to a gas.
Exothermic reaction	an exothermic reaction is a chemical reaction during which heat energy is released in to the surroundings.
Filtrate	is the liquid that passes through the filter paper during the physical separation technique of filtration. The liquid is usually clear.
Filtration	is a method of physically separating solids from a liquid by passing the mixture through a porous material such as filter paper or glass wool.
Forward reaction	the forward reaction is the direction of the chemical reaction when read from left to right, in an equilibrium reaction equation.
Fossil fuels	are formed over millions of years from the remains of ancient buried organisms. Examples include oil and coal.
Fractional distillation	is a method of separating a mixture of liquids with different boiling points using distillation.

Freezing point	is the temperature at which all of a liquid changes into a solid.
Functional group	a functional group is an atom or group of atoms that give an organic compound its characteristic chemical properties.
Gas	is a state of matter in which the particles in have no fixed shape or volume. Particles of a gas are far apart, can be compressed, experience no attractive forces between the molecules and move randomly and rapidly in their surroundings.
Global warming	is the gradual increase in the average global temperature.
Graphene	is a covalent network solid that consists of a single-layer sheet of carbon atoms arranged in hexagonal rings. The structure is one atom thick.
Greenhouse effect	the greenhouse effect is the trapping of heat energy radiated from the Earth's surface by greenhouse gases present in the atmosphere.
Greenhouse gases	are the gases in the atmosphere which absorb infrared radiation, causing an increase in air temperature.
Group	a group is a vertical column of elements in the periodic table.
Halogens	are the elements in group 17 of the periodic table. These non-metals have seven valence electrons in their outermost shell.
Heat energy	is the energy transferred from a hotter to a cooler part of a system, due to the temperature gradient.
Heterogeneous mixture	a heterogeneous mixture is a mixture with an uneven distribution of constituents resulting in a non-uniform composition. The physical and chemical properties vary throughout the mixture.
Homogeneous mixture	a homogeneous mixture is a mixture with an even distribution of constituents resulting in a uniform composition throughout and therefore has uniform properties.
Homologous series	a homologous series is a series of related organic compounds with the same functional group.
Hydrocarbons	are organic compounds that contain only carbon and hydrogen atoms.
Hydrogen bonding	is the strong intermolecular force of attraction between certain molecules that contain H–F, O–H or N–H bonds. Water molecules experience hydrogen bonding.
Hypothesis	a hypothesis is an explanation of an observation or phenomenon which must be tested by experimental investigations.
Immiscible	describes two or more liquids that will not mix together to form a homogeneous mixture (for example, oil and water).
Incomplete combustion	is a combustion reaction where the supply of oxygen to the reactants is insufficient. Possible products include carbon monoxide and carbon.
Indicator	is an organic dye that changes color, reversibly, over a narrow pH range in an acid–base reaction, according to whether a solution is acidic or alkaline.
Inductive reasoning	is where our thoughts move from the specific to general statements, based on evidence collected through observations during experiments.

Inference	is a conclusion that is based on experimental evidence and observations.
Insulator	an insulator is a material which does not readily allow the passage of electrons through the material.
Intermolecular forces	are forces of attraction between molecules.
Ion	an ion is a charged particle formed when an atom (or group of atoms) gains or loses one or more electrons.
Ionic bond	is the electrostatic attraction between the charges of a positively charged cation (formed by metals) and a negatively charged anion (formed by non-metals).
Ionization energy	is the minimum energy required to remove an electron from the outermost electron shell of a gaseous atom.
Isoelectronic	is the term used to describe elements and ions that have identical electron configurations.
Isolated system	an isolated system is a system in which no matter or energy can enter or leave the system.
Isomer	an isomer is a compound with a different structural formula to another compound, but an identical chemical formula.
Isotopes	are atoms of the same element with the same number of protons and electrons but a different number of neutrons. The mass numbers of isotopes are different.
Kelvin	is the SI unit of temperature.
Kinetic energy	is the form of energy an object or particle has as a result of its motion.
Kinetic theory of gases	is a simple model that describes the movement of gases. It states that molecules of a gas are in constant, random motion.
Le Chatelier's principle	states that if a change is made to a system that is in equilibrium, the balance that exists between the forward and reverse reactions will shift to return the system to equilibrium.
Limiting reagent	is the reactant that is completely consumed in a chemical reaction. The number of moles of a limiting reagent determines the amount of product formed.
Liquid	is a state of matter in which the particles have a fixed volume, but no fixed shape. It cannot be compressed as the particles are close together, vibrate and have a limited amount of movement.
Lone pair of electrons	a lone pair of electrons is a pair of electrons that are not involved in the formation of covalent bonds.
Malleability	is the ability of a metal to be pounded into a sheet or any other shape without shattering or breaking.
Mass number	is the total number of protons and neutrons found in the nucleus of an atom.
Matter	is made up of particles such as atoms, occupies a place in space, is in constant motion, and has a mass.
Melting point	is the temperature at which a solid completely changes into a liquid.

Metals
are a class of chemical elements that are composed of a lattice of positive ions surrounded by a sea of delocalized electrons. Metals form positive ions (cations) when they react to form compounds.

Metallic bond
a metallic bond is the electrostatic attraction between a lattice of positively charged metal ions and delocalized electrons.

Miscible
describes two or more liquids that will diffuse together and form a single phase (for example, alcohol and water). The components form a homogeneous mixture.

Mixture
a mixture is a combination of two or more substances that are not chemically bonded and can be separated using physical techniques such as filtration, crystallization, distillation and chromatography.

Models
are used by scientists to explain phenomena that may not be able to be observed directly.

Molar mass
is the mass of one mole of a substance.

Mole
the mole is the SI unit that describes a fixed number of particles of a substance.

Negative charge
is the result of an element having a larger number of electrons than protons.

Neutral charge
is the result of an element having an equal number of electrons and protons.

Neutralization
is the type of chemical reaction between an acid and a base to form a salt and water.

Neutron
is a subatomic particle which is found in the nucleus of atoms. It has no charge.

Noble gases
are elements found in Group 18 of the periodic table. They have a full outer electron shell and are unreactive.

Non-directional bonding
is bonding with no directional character. This means bonding interactions can occur in more than one direction from a given atom. Ionic and metallic bonding are non-directional.

Non-metals
are elements that are poor conductors of heat and electricity, gain electrons when involved in a chemical reaction and form ionic compounds with metals.

Nucleus
a nucleus is the very small central core of an atom, containing most of the atomic mass in the form of protons and neutrons.

Open system
an open system is a system in which both materials and energy can enter or leave the system.

Orbitals
are regions surrounding the nucleus of an atom where there is a 95% probability of finding an electron.

Organic compound
an organic compound is a carbon-based compound in which carbon atoms are covalently bonded to each other, as well as other elements such as hydrogen, oxygen and nitrogen.

Oxidation
is a chemical reaction involving the gain of oxygen or the loss of electrons.

Oxidizing agent	an oxidizing agent is a substance that oxidizes another species by making it lose electrons, and is itself reduced by gaining these electrons.
Paper chromatography	is a physical separation technique designed to separate the components of mixtures, such as dyes and pigments. The mixture is dissolved in the solvent (or mobile phase). The mobile phase then moves up the chromatography paper (or stationary phase), separating out the components.
Period	a period is a horizontal row of elements on the periodic table.
Periodic table	the periodic table is an arrangement of elements in order of increasing atomic number.
pH scale	the pH scale is a logarithmic number scale (0 to 14) which represents the concentration of hydrogen ions in a solution. It is used to describe the strength of an acid or an alkali.
Physical change	a physical change is one where matter undergoes a change in state.
Physical properties	of an element such as density, melting point and electrical conductivity can be observed without changing the chemical composition of the substance.
Polyatomic ion	is a group of two or more atoms chemically bonded together with an overall charge. For example, the carbonate ion $CO_3{}^{2-}$.
Polymer	is a macromolecule that is made up of many repeating units called monomers.
Positive charge	is the result of an atom of an element having a larger number of protons than electrons.
Precision	is a measure of the variation in the results of identical trials of an experiment. If there is less variation in the range of results, the value may be expressed with a larger number of significant figures and may be more precise.
Products	are the chemical elements or compounds that are formed during a chemical reaction. They are found on the right hand side of a chemical equation.
Proton	a proton is a positively charged subatomic particle which is found in the nucleus of an atom.
Precipitate	a precipitate is an insoluble solid formed during a chemical reaction between two soluble ionic substances. A precipitate sometimes forms during single and double replacement reactions.
Qualitative data	includes non-numerical information that is obtained through observations rather than measurements.
Quantitative data	is numerical information that is obtained by measurements during a chemical reaction.
Random error	are the result of uncontrolled variables in experiments. They cannot be eliminated but they can be reduced by conducting repeated trials.
Rate of reaction	is a numerical description of the rate at which reactants are converted into products.

Reactants are the chemical elements or compounds that are transformed into products during a chemical reaction. They are found on the left hand side of a chemical equation.

Redox reactions are two interconnected chemical reactions that must occur together. Separately they are known as oxidation and reduction reactions.

Reducing agent a reducing agent is a substance that reduces another species by making it gain electrons, and is itself oxidized by losing these electrons.

Reduction is a chemical reaction involving the loss of oxygen or the gain of electrons.

Reflux is a laboratory technique that involves the continuous evaporation and condensing of a reaction mixture solvent.

Relative atomic mass is the weighted average of the atomic masses of the isotopes of an element and their relative abundance.

Residue is the insoluble solid element or compound separated by the filter paper during filtration.

Rusting is the corrosion or oxidation of pure iron or steel to form iron(III) oxide, Fe_2O_3.

Salts are chemical compounds formed when a metal cation bonds with a non-metal anion. Common salts include sodium chloride, potassium bromide, and calcium nitrate.

Salt bridge a salt bridge is a piece of apparatus (normally glass) that is filled with an electrolyte and is part of an electrochemical cell called a voltaic cell.

Saturated solution a saturated solution is one which will not dissolve any more solute at a particular temperature.

Scientific method is an experiment-based approach designed to test hypotheses. It generally follows the steps of making observations, generating questions, forming hypotheses, testing the hypotheses, collecting and analyzing data, drawing conclusions and modifying the original experimental approach.

Significant figures the number of significant figures in a value is the number of digits which represent the precision of a given measurement.

Single covalent bond a single covalent bond is formed by the electrostatic attraction between a single pair of shared electrons in the nuclei of the atoms.

Solid is a state of matter in which the particles have a fixed volume and shape and cannot be compressed, as the particles in a solid are very close together. The attractive forces between the particles hold them in place to form a regular, close packed arrangement. The particles vibrate, but do not move.

Solubility is a measure of the amount of a solute that will dissolve in a given volume of solvent at a specific temperature.

Solute a solute is the solid substance which dissolves in the solvent to form a solution.

Solution a solution is a homogeneous mixture in which the particles of solute and solvent are evenly distributed.

Solvation	is the interaction between a solute and a solvent that results in the solute becoming soluble in the solvent.
Solvent	a solvent is a substance, usually a liquid, that dissolves a solute to form a solution. The solvent can be polar (for example, water) or non-polar (for example, hexane)
Specific heat capacity	is the amount of heat energy required to raise the temperature of 1 g of a pure substance by 1°C.
Spontaneous reaction	a spontaneous reaction is a reaction that reaches completion or equilibrium under a given set of conditions, without the need for any external intervention.
Standard temperature and pressure	are reaction conditions with a temperature of 0°C (273 K) and a pressure of 1 atmosphere (101,325 pascals).
State symbols	include solid (s), liquid (l), gas (g) and aqueous (aq). Aqueous means a solute is dissolved in water.
Stoichiometry	is the quantitative relationship between reactants and products in a balanced chemical equation.
Sublimation	is the direct change of state from a solid to a gas on heating.
Supersaturated solution	is a solution that contains more solute than required to form a saturated solution, for a given volume and temperature of a solvent.
Suspension	a suspension is a mixture of small, insoluble solid particles that are dispersed in a gas or liquid.
Synthesis reaction	is the formation of a single product from two or more simpler reactants.
Systematic errors	are the result of the experimental procedure or instrumentation being used.
Temperature	is a measure of how hot or cold matter is. It is indicative of the average kinetic energy of the particles.
Thermochemistry	is the study of heat changes that occur during chemical reactions.
Thermodynamics	is the study of energy interconversions in chemical reactions.
Titration	is a quantitative, analytical technique used to determine the unknown concentration of a reactant solution. It is typically used during acid–base neutralization reactions.
Transition elements	are found in groups 3–12 of the periodic table. Their properties include the formation of colorful complexes.
Triple covalent bond	a triple covalent bond is the sharing of three pairs of electrons between two bonding atoms.
Ultraviolet light	is a form of electromagnetic radiation. In the stratosphere, 95% of this form of harmful light is absorbed by the ozone layer.
Unsaturated hydrocarbons	are compounds that contain at least one carbon–carbon double covalent bond.
Valence electron	a valence electron is an electron found in the outermost electron shell of an atom.

Visible light   is a form of electromagnetic radiation to which the human eye can respond.

Vaporization   is the change in state from a liquid to a gas.

Variable   in an experiment, a variable is a parameter that can change between different trials. Variables are usually classified as the independent, dependent and control variables.

Voltaic cell   a voltaic cell is any deice that produces an electromotive force (emf) by converting chemical to electrical energy.

Volume   is the amount of 3-dimensional space occupied by a solid, liquid or gas. The standard units of measurement for volume in science are cm^3 and dm^3.

Volumetric analysis   is a quantitative analysis technique that involves the accurate measurement of volumes of liquids of known concentrations, to determine the unknown concentration of a solution. An example of this technique is an acid–base titration.

Weak acid   weak acids such as carbonic acid only partly ionize in water, giving a small concentration of hydrogen ions.

Index

The entries in **bold** are explained in the glossary.

The publisher would like to thank the following for permissions to use their photographs:

Cover: Alamy Stock Photo; **p2(T)**: Magicinfoto/Shutterstock; **p2(B)**: Nagel Photography/Shutterstock; **p3(T)**: Yuri Arcurs/Shutterstock; **p3(B)**: JPL/NASA; **p4**: Imagevixen/Shutterstock; **p5**: Bill Bachman/Science Photo Library; **p7**: Granger Historical Picture Archive/Alamy Stock Photo; **p8**: © A. Stodolna, M.J.J. Vrakking and co-workers; **p12**: Specta/Shutterstock; **p13(B)**: Anatoli Styf/Shutterstock; **p16**: Yasni/Shutterstock; **p14**: Paul Fearn/Alamy Stock Photo; **p18**: Blazej Lyjak/Shutterstock; **p20**. Sciencephotos/Alamy Stock Photo; **p26 (L)**: Everett Historical/Shutterstock; **p26 (R)**: GIANNI TORTOLI/SCIENCE PHOTO LIBRARY; **p27 (T)**: Suwin/Shutterstock; **p27 (B)**: Jack Young - Places/Alamy Stock Photo; **p29 (R)**: MARTYN F. CHILLMAID/SCIENCE PHOTO LIBRARY; **p29 (L)**: Anyaivanova/Shutterstock; **p28**: The Biochemist Artist/Shutterstock; **p30 (T)**: GIPhotoStock/SCIENCE PHOTO LIBRARY; **p30 (B)**: Smereka/Shutterstock; **p31**: Fotokostic/Shutterstock; **p35**: Alexandre Dotta/SCIENCE PHOTO LIBRARY; **p36**: GIPhotoStock/SCIENCE PHOTO LIBRARY; **p37 (T)**: ANDREW LAMBERT PHOTOGRAPHY/SCIENCE PHOTO LIBRARY; **p37 (B)**: Wikipedia; **p38 (C)**: GONUL KOKAL/Shutterstock; **p38 (B)**: Urfin/Shutterstock; **p39 (T)**: Kim Christensen/Shutterstock; **p39 (C)**: Phil Degginger/SCIENCE PHOTO LIBRARY; **p39 (B)**: Anucha Sirivisansuwan/Shutterstock; **p42**: Molekuul.be/Shutterstock; **p46**: Igor Kisselev/Shutterstock; **p58**: Jeffrey Kargel, USGS/NASA JPL/AGU; **p60 (T)**: STEVE GSCHMEISSNER/SPL/Getty Images; : Ian Bottle/Alamy Stock Photo; **p60 (B)**: A Katz/Shutterstock; **p61 (T)**: Alpha and Omega Collection/Alamy Stock Photo; **p61 (B)**: TORWAISTUDIO/Shutterstock; **p62**: Bill Bachman/Alamy Stock Photo; **p67**: Irina Sen/Shutterstock; **p69**: Feng Yu/Shutterstock; **p71**: HP Canada/Alamy Stock Photo; **p72**: Science Photo Library/Alamy Stock Photo; **p75**: CHARLES D. WINTERS/SCIENCE PHOTO LIBRARY; **p77**: Everett Historical/Shutterstock; **p78 (T)**: Richard Ellis/Alamy Stock Photo; **p78 (T)**: Marbury/Shutterstock; **p80**: DAVID HAY JONES/SCIENCE PHOTO LIBRARY; **p84 (T)**: ANDY MURCH/VISUALS UNLIMITED, INC. /SCIENCE PHOTO LIBRARY; **p84 (B)**: SPUTNIK/SCIENCE PHOTO LIBRARY; **p85 (T)**: Alslutsky/Shutterstock; **p85 (B)**: Photodisc/Getty Images; **p87 (T)**: ALEX BARTEL/SCIENCE PHOTO LIBRARY; **p87 (B)**: YVES SOULABAILLE/LOOK AT SCIENCES/SCIENCE PHOTO LIBRARY; **p86**: Rynio Productions/Shutterstock; **p88 (T)**: SCIENCE PHOTO LIBRARY; **p88 (B)**: Reidl/Shutterstock; **p90**: Todor Dobrev/Shutterstock; **p91 (TL)**: Ulga/Shutterstock; **p91 (TR)**: PUVIL/Shutterstock; **p91 (CL)**: Tanapat Mutitakun/Shutterstock; **p91 (CR)**: Artography/Shutterstock; **p92**: SINITAR/Shutterstock; **p93**: SCIENCE PHOTO LIBRARY; **p95**: ANDREW LAMBERT PHOTOGRAPHY/SCIENCE PHOTO LIBRARY; **p108 (T)**: Dallas Reeves/Shutterstock; **p109 (T)**: D.O.Hill/Wikipedia; **p108 (B)**: Norhayati/Shutterstock; **p109 (B)**: ASHLEY COOPER/SCIENCE PHOTO LIBRARY; **p110**: M. Behrens et al., Science 336, 893 (2012); p111: Prisma by Dukas Presseagentur GmbH/Alamy Stock Photo; **p113**: © Berkeley Lab; **p116**: Andrii Gorulko/Shutterstock; **p118**: Anucha Cheechang/Shutterstock; **p119**: Levgeniia Miroshnichenko/Shutterstock; **p120**: Lakov Filimonov/Shutterstock; **p124**: Aleksey Klints/Shutterstock; **p134 (T)**: Kavuto/iStockphoto; **p134 (B)**: Vietnam Stock Images/Shutterstock; **p135 (T)**: Photodisc/Getty Images; **p135 (B)**: Ambelrip/Shutterstock; **p137 (BL)**: Love Silhouette/Shutterstock; **p136**: EyeSeeMicrostock/Shutterstock; **p137 (BC)**: Ggw1962/Shutterstock; **p137 (BR)**: Jiri Vaclavek/Shutterstock; **p137 (B)**: GIPhotoStock/SCIENCE PHOTO LIBRARY; **p137 (T)**: MARTYN F. CHILLMAID/SCIENCE PHOTO LIBRARY; **p140**: GIPhotoStock/SCIENCE PHOTO LIBRARY; **p141 (TR)**: Niyazz/Shutterstock; **p141 (TL)**: CLIVE FREEMAN/BIOSYM TECHNOLOGIES/SCIENCE PHOTO LIBRARY; **p141 (TCL)**: CLIVE FREEMAN/BIOSYM TECHNOLOGIES/SCIENCE PHOTO LIBRARY; **p142**: Natursports/Shutterstock; **p144 (BL)**: Jorge Salcedo/Shutterstock; **p144 (BR)**: Nature Picture Library/Alamy Stock Photo; **p145 (TR)**: KPG_Payless/Shutterstock; **p145 (CL)**: GIPhotoStock/SCIENCE PHOTO LIBRARY; **p146 (T)**: Laura Hennebery/Thomson Digital and Pantek Arts Ltd; **p146 (B)**: MARTYN F. CHILLMAID/SCIENCE PHOTO LIBRARY; **p147 (L)**: Photodisc/Getty Images; **p147 (R)**: Richard Hutchings/SCIENCE PHOTO LIBRARY; **p148**: Mikeledray/Shutterstock; **p149**: SCIENCE PHOTO LIBRARY; **p150 (TL)**: Ggw/Shutterstock; **p150 (BL)**: MARTYN F. CHILLMAID/SCIENCE PHOTO LIBRARY; **p150 (TR)**: Hurst Photo/Shutterstock; **p150 (BR)**: KTSDESIGN/SCIENCE PHOTO LIBRARY; **p151**: Vlue/Shutterstock; **p154**: Alexandre Dotta/SCIENCE PHOTO LIBRARY; **p155**: PAUL RAPSON/SCIENCE PHOTO LIBRARY; **p162 (T)**: Inger Eriksen/Shutterstock; **p162 (BL)**: Alexey Stiop/Shutterstock; **p162 (BR)**: WUT.ANUNAI/Shutterstock; **p163 (T)**: Michael Fitzsimmons/Shutterstock; **p164**: Hitendra Sinkar/Alamy Stock Photo; **p163 (B)**: Nick Henn/Shutterstock; **p165 (B)**: SCIENCE PHOTO LIBRARY; **p165 (T)**: GEORGE BERNARD/SCIENCE PHOTO LIBRARY; **p165 (C)**: MARTYN F. CHILLMAID/SCIENCE PHOTO LIBRARY; **p166**: James Clarke/Shutterstock; **p167**: GIPhotoStock/SCIENCE PHOTO LIBRARY; **p173 (R)**: Vincent Noel/Shutterstock; **p173 (L)**: Tigergallery/Shutterstock; **p177**: COLLECTION ABECASIS/SCIENCE PHOTO LIBRARY; **p178**: Satishyewlekar/Shutterstock; **p179**: Charles D. Winters/SCIENCE PHOTO LIBRARY; **p190 (T)**: Matthew J Thomas/Shutterstock; *Continued on last page* **p190 (B)**: Greg Amptman/Shutterstock; **p191 (T)**: GeorginaCaptures/Shutterstock; **p191 (B)**: C. Ortiz Rojas/ational Oceanic and Atmospheric Administration/Department of Commerce; **p192**: Prasit Rodphan/Shutterstock; **p193 (T)**:

MARTIN, CUSTOM MEDICAL STOCK PHOTO/SCIENCE PHOTO LIBRARY; **p193 (C)**: Inu/Shutterstock; **p193 (B)**: Testing/Shutterstock; **p194 (CL)**: Philip Date/Shutterstock; **194 (CR)**: Sezer66/Shutterstock; **p194 (BL)**: Dundanim/Shutterstock; **194 (BR)**: Jan Lipina/Shutterstock; **p195**: Chiara Casanova/Shutterstock; **p198**: 3523studio/Shutterstock; **p199**: Balhash/iStockphoto; **p201 (L)**: SCIENCE PHOTO LIBRARY; **p201 (R)**: SCIENCE PHOTO LIBRARY; **p206**: GIPhotoStock/SCIENCE PHOTO LIBRARY; **p212 (T)**: SSSCCC/Shutterstock; **p213 (B)**: GARY BROWN/SCIENCE PHOTO LIBRARY; **p212 (B)**: ©Zaha Hadid Architects; **p215 (T)**: William Putman/NASA Goddard Space Flight Center/SCIENCE PHOTO LIBRARY; **p214**: ORNL/SCIENCE PHOTO LIBRARY; **p215 (T)**: ROYAL INSTITUTION OF GREAT BRITAIN / SCIENCE PHOTO LIBRARY; **p215 (B)**: Igor Petrushenko/Shutterstock; **p218**: ALFRED PASIEKA/SCIENCE PHOTO LIBRARY; **p220**: Swiss Studio/Shutterstock; **p222**: Stockagogo, Craig Barhorst/Shutterstock; **p227**: PRILL/Shutterstock **p229**: Benjah-bmm27/Wikipedia; **p231**: MEHAU KULYK/SCIENCE PHOTO LIBRARY; **p232 (L)**: Juriaan Wossink/Shutterstock; **p232 (C)**: Molekuul_be/Shutterstock; **p232 (R)**: Kim Christensen/Shutterstock; **p236**: Magnetix/Shutterstock; **p237**: Katharina M/Shutterstock; **p238**: Celiafoto/Shutterstock; **p239**: Hammad/Q2A Media; **p241**: First Class Photography/Shutterstock; **p246**: NASA/SCIENCE PHOTO LIBRARY; **p247 (T)**: Gritsalak Karalak/Shutterstock; **p247 (C)**: John Mainstone, University of Queensland/Wikipedia; **p247 (B)**: Siim Sepp/Shutterstock; **p248**: Charles D. Winters/SCIENCE PHOTO LIBRARY; **p249**: Willyam Bradberry/Shutterstock; **p250**: Alexandre Dotta/SCIENCE PHOTO LIBRARY; **p252**: DAVID GUYON, THE BOC GROUP PLC/ SCIENCE PHOTO LIBRARY; **p257**: Wearset Ltd, and HL Studios; **p263**: TREVOR CLIFFORD PHOTOGRAPHY/SCIENCE PHOTO LIBRARY; **p271**: CLAUDE NURIDSANY & MARIE PERENNOU/SCIENCE PHOTO LIBRARY; **p269**: Giedre Vaitekune/Shutterstock; **p274 (T)**: ART Collection/Alamy Stock Photo; **p274 (B)**: Granger Historical Picture Archive/Alamy Stock Photo; **p275**: Gorodenkoff/Shutterstock; **p276 (T)**: Funny Solution Studio/Shutterstock; **p276 (B)**: MMPOP/Shutterstock; **p277 (T)**: Chris Howey/Shutterstock; **p277 (B)**: Ibreakstock/Shutterstock; **p279**: MIKKEL JUUL JENSEN/SCIENCE PHOTO LIBRARY; **p278**: GIPhotoStock/SCIENCE PHOTO LIBRARY; **p282**: Tony Hunter; **p281 (R)**: Jeff Bowles, Roger Courthold, Mike Ogden, Jeff Edwards, Russell Walker, Clive Goodyer, Jamie Sneddon, Q2A and Tech Graphics; **p281 (L)**: H. S. Photos/SCIENCE PHOTO LIBRARY; **p284**: Hammad/Q2A Media; **p289**: Dinodia Photos/Alamy Stock Photo; **p293 (T)**: ANDREW LAMBERT PHOTOGRAPHY/SCIENCE PHOTO LIBRARY; **p293 (B)**: ANDREW LAMBERT PHOTOGRAPHY/SCIENCE PHOTO LIBRARY; **p304 (T)**: Jody Ann/Shutterstock; **p304 (B)**: Aleksandar Mijatovic/Shutterstock; **p305 (T)**: Benny Marty/Shutterstock; **p305 (B)**: RichardBakerHeathrow/Alamy Stock Photo; **p306**: GoneWithTheWind/Shutterstock; **p307 (T)**: Hagai Nativ/Alamy Stock Photo; **p307 (B)**: CHARLES D. WINTERS/SCIENCE PHOTO LIBRARY; **p309**: GIPhotoStock/SCIENCE PHOTO LIBRARY; **p311**: Manjay/Q2A Media; **p314**: ANDREW LAMBERT PHOTOGRAPHY/SCIENCE PHOTO LIBRARY; **p318 (T)**: Photong/Shutterstock; **p318 (B)**: Petarg/Shutterstock; **p320 (T)**: Gareth Boden; **p320 (CT)**: Ingram/Alamy Stock Photo; **p320 (CB)**: Valentina Razumova/Shutterstock; **p320 (B)**: Maks Narodenko/Shutterstock.

Artwork by Aptara Corp. and OUP.